The American Association of Variable Star Observers

AAVSO

Annual Report 2008–2009

The American Association of Variable Star Observers
49 Bay State Road
Cambridge, MA 02138
USA

Telephone: 617-354-0484 Fax: 617-354-0665

email: aavso@aavso.org website: http://www.aavso.org

Annual Report Website:
http://www.aavso.org/aavso/annualreports.shtml

On the cover...

Tricolor image of the Pleiades created by Alson Wong from deep images taken by Tom Krajci during the early commissioning of the Bright Star Monitor.

When epsilon Aurigae became an interesting target last year, it became obvious that none of the telescopes the AAVSO had access to were capable of obtaining photometry on such a bright object. We also realized that this was not an isolated case; there are hundreds of bright variables that are being overlooked by professionals and amateurs alike.

We looked at what would be required to do the job, and decided to use a scientific CCD camera, a filter wheel with filters, a GOTO mount, and commercial software, and to place it at a good site. We dubbed the system the Bright Star Monitor (BSM). The main system parameters are: Telescope, Takahashi FS-60CB with field flattener, 60mm f/6.2; Camera, SBIG ST-8XME; Filterwheel, SBIG CFW-9; Filters, Johnson/Cousins BVRcIc plus Clear; Mount, Celestron CGEM; Location, Astrokolkhoz Observatory (Tom Krajci, Cloudcroft, NM), Latitude +32:52:31, Longitude +105:47:11; Elevation, 9440ft/2880m; Software: MaximDL, ACP, ACP Scheduler.

Picture credits

In additon to images from the AAVSO and its archives, the editors gratefully acknowledge the following for their image contributions: Mary Glennon, Mario Motta, NASA, Gary Poyner, Msgr. Ronald Royer, the Mary Lea Shane Archives of the Lick Observatory, and Wheatley, et al. *2003,* MNRAS, *345, 49.*

Table of Contents

1. About the AAVSO

Participants in the AAVSO's 98th Annual Meeting, 2009

AAVSO Vision

The AAVSO seeks to be the world-recognized leader in information and data on variable stars.

The AAVSO's Mission

The AAVSO is an international non-profit organization of variable star observers whose mission is:

- to observe and analyze variable stars
- to collect and archive observations for worldwide access
- to forge strong collaborations between amateur and professional astronomers
- to promote scientific research and education using variable star data.

About the AAVSO

The American Association of Variable Star Observers (AAVSO) is a non-profit worldwide scientific and educational organization of amateur and professional astronomers who are interested in stars that change in brightness—variable stars.

The AAVSO was founded in 1911 to coordinate variable star observations—made largely by amateur astronomers—for Harvard College Observatory. The AAVSO was incorporated in the Commonwealth of Massachusetts in 1918 as a non-profit scientific and educational organization. Today, as an independent, private research organization headquartered in Cambridge, Massachusetts, with members and observers in 52 countries, and an archive of over 17.5 million variable star observations, it is the world's largest association of variable star observers.

Membership in the AAVSO is open to anyone—professionals, amateurs, and educators alike—interested in variable stars and in contributing to the support of valuable research.

Professional astronomers have neither the time nor the telescopes needed to gather data on the brightness changes of thousands of variables, and amateurs make a real and useful contribution to science by observing variable stars and submitting their observations to the AAVSO International Database.

What We Do

The AAVSO coordinates, evaluates, compiles, processes, publishes, and disseminates variable star observations to the astronomical community throughout the world.

Observers send their data to Headquarters, where they are checked, processed, and added to the AAVSO International Database. The AAVSO and its observers frequently provide the professional community with archival data, intensive monitoring of interesting variable stars, and target-of-opportunity event notification for coordinated observing campaigns and satellite observations.

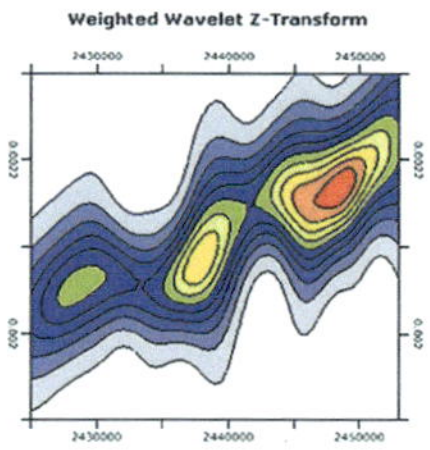

AAVSO publications provide the astronomical community with valuable information. The type of published information is diverse, and includes *The Journal of the AAVSO*, a peer-reviewed collection of scientific papers focused on variable stars, the *Manual for Visual Observing*, now available in eight languages, the *CCD Observing Manual*, the quarterly *AAVSO Newsletter*, the *Eclipsing Binary and RR Lyrae Ephemerides*, and the *AAVSO Annual Report*.

Additionally, the AAVSO is actively involved in education and outreach. We have several programs designed to assist with disseminating information to educators and the public.

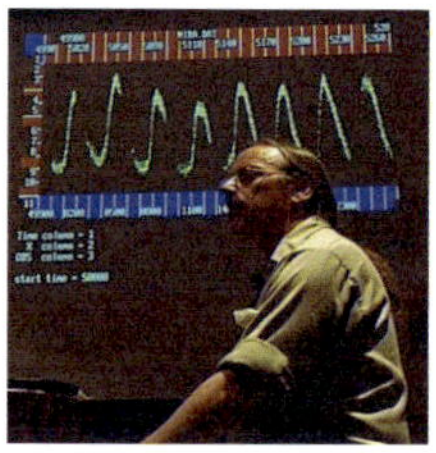

The AAVSO has an active Mentor Program that is available to any observer requesting personal instruction in observing techniques and methods.

The Speakers Bureau is a service established for people and groups looking for enthusiastic, knowledgeable speakers.

Our Presentation Library offers free POWERPOINT™ presentations on variable stars, observing techniques, and other astronomical topics.

Our Writers Bureau offers variable star and topical astronomy content on a monthly basis to editors of astronomy club and society newsletters.

Variable Star Astronomy (VSA) is a flexible set of hands-on educational materials, activities, and investigations, based on the AAVSO's unique electronic database of variable star measurements.

Members and observers have a unique opportunity to present and exchange ideas at the AAVSO meetings. The AAVSO organizes two meetings a year, one in the fall and one in the spring. The fall meeting is the official AAVSO annual meeting that is always held at or near the AAVSO Headquarters in Cambridge, MA. The spring meeting is held outside of the state of Massachusetts with the intention of attracting more members and observers to attend. Everyone interested in the AAVSO and its activities is invited and encouraged to participate in these exciting events.

What Are Variable Stars?

Variable stars are stars that change brightness. The brightness changes of these stars can range from a thousandth of a magnitude to as much as twenty magnitudes over periods of a fraction of a second to years, depending on the type of variable star. Over 100,000 variable stars are known and catalogued, and many thousands more are suspected to be variable.

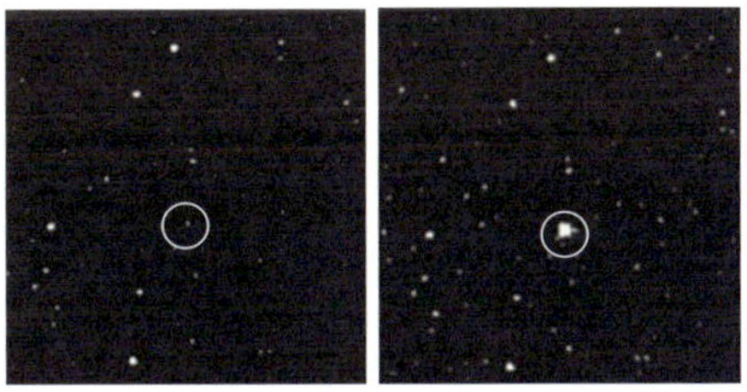

The variable star U Geminorum in its faint state (left) and its bright state (right)

There are a number of reasons why variable stars change their brightness. Pulsating variables, for example, swell and shrink due to internal forces. An eclipsing binary will dim when it is eclipsed by a faint companion, and then brighten when the occulting star moves out of the way. Some variable stars are actually extremely close pairs of stars, exchanging mass as one star strips the atmosphere from the other.

The different causes of light variation in variable stars provide the impetus for classifying the stars into different categories. Variable stars are classified as either intrinsic, wherein variability is caused by physical changes such as pulsation or eruption in the star or stellar system, or extrinsic, wherein variability is caused by the eclipse of one star by another, the transit of an extrasolar planet, or by the effects of stellar rotation.

Why Observe Variable Stars?

Variable stars need to be systematically observed over decades in order to determine their long-time behavior. Professional astronomers have neither the available time nor the unlimited telescope access needed to gather data on the brightness changes of thousands of variable stars. Thus, it is amateur astronomers utilizing visual, photographic, photoelectric, and CCD techniques who are making a real and highly useful contribution to science by observing variable stars and submitting their observations to the AAVSO International Database. These important data are needed to analyze variable star behavior, to schedule satellite observations of certain stars, to correlate data from satellite and ground-based observations, and to make computerized theoretical models of variable stars possible.

Research on variable stars is important because it provides information about stellar properties, such as mass, radius, luminosity, temperature, internal and external structure, composition, and evolution. Some of this information would be difficult or impossible to obtain any other way. In many cases, it is the nature of the variability that provides the clues to the answers. This information can then be used to understand other stars.

Variable stars continue to play a crucial role in our understanding of the universe. Cepheid variables have played a major part in determining distances to far-away galaxies and determining the age of the Universe. Mira variables give us a glimpse into the future evolution of our own star, the Sun. Accretion disks in cataclysmic variables help us to understand larger scale disk behavior, like the activity inside active galaxies with super-massive black holes. Supernovae have led us to the surprising realization that the expansion of the Universe is accelerating. Even the search for extra-terrestrial life is illuminated by variable stars. Transiting extrasolar planets provide clues into the processes of planetary formation, and the very stuff life as we know it is made of comes from the hearts of stars that explode in the final stages of their evolution.

The AAVSO International Database

The AAVSO International Database has over 17.5 million variable star brightness estimates going back over one hundred years. It is the largest and most comprehensive digital variable star database in the world. Over 1,000,000 new variable star brightness measurements are added to the database every year by over 700 observers from all over the world.

Quality

The AAVSO International Database is not only the largest but also the highest quality database available to researchers. The AAVSO and its technical staff spend more time and resources on database maintenance and quality control than any other organization.

Quality control begins before the observation is even made. Extensive training materials are sent to new AAVSO observers and a large section of the AAVSO website is designed specifically for observing techniques. The AAVSO holds two meetings per year where members come together to discuss their observing strategies, compare results, and much more. Workshops are routinely held at these meetings, bringing the best professionals in the field in contact with the observers. Since 2000, workshops have been held on CCD imaging, Eclipsing Binary star observing, GRB afterglow hunting, data mining, and data analysis. The AAVSO also has an active mentoring program for new observers.

We have data entry error checks at every stage in the process. Whether the observer is using WebObs, PCObs, or sending their data in via e-mail, we have error checking routines running to automatically identify the most common data entry errors. In addition, every month we comb through all the observations using both human scrutiny and automated programs to look for misidentifications, typos, and any other errors. The best check, however, is the observers themselves who check their submitted data by using the Quick Look file, Light Curve Generator, and by comparing their own records with their observations in the AAVSO International Database.

Data validation ensures the quality of our permanent archives

This practice is what really separates AAVSO data from others. Every data point that comes from the AAVSO International Database has been validated—that is, put through a rigorous system of data integrity checks. This system involves running automated programs and also requires a human being to actually look at and validate each data point. Not a point gets through the system without being looked at by a real person. This combination of techniques takes advantage of the benefits that both humans and automation can bring to the process, and it is applied not only to new observations, but also to every observation in the database, even the ones made a century ago.

Observers

The AAVSO International Database would not exist without the dedication, tireless effort, and enthusiasm of thousands of variable star observers. Our observers come from all over the world. Over two-thirds of AAVSO observers contributing data come from outside of the United States. Thanks to this broad network of observers we have coverage across most time zones and latitudes regardless of weather or other regional disruptions. To make it easier for the widely-scattered AAVSO members and observers to gather together in person, the AAVSO meetings held every spring or summer take place in different parts of the United States or, as often as possible, in different countries.

Mary Glennon, AAVSO member-observer since 1999

The AAVSO receives observations from members of other variable star observing associations around the world for inclusion in the AAVSO International Database and dissemination to the astronomical community worldwide. These observations are sent regularly by the group leader/representative or directly by the group members themselves. The AAVSO values these fruitful, mutually beneficial collaborations, and truly appreciates the ongoing efforts of everyone involved in working together for the benefit of the astronomical community.

Access

Observations from the database are available to anyone at any time. For raw observations, simply fill out our online request form. For access to light curves, use our Light Curve Generator which works in all browsers (you do not need JAVA or any special plug-ins), and for really quick access to recent data, visit our Quick Look file. Our online systems are updated every ten minutes with the latest data.

Observing Variable Stars

Astronomy is a unique science that cannot be studied in a typical laboratory setting here on earth. Instead, astronomers turn their attention and telescopes to the sky in order to study their subjects. Since professional astronomers often do not have the telescope time needed to follow a particular star or group of stars, the dedication of amateur astronomers is often an invaluable means of collecting information. Nowhere is this more true than in the field of variable star astronomy. Since 1911, thousands of amateur astronomers from all over the world and from all backgrounds have contributed observations, one at a time, to make up the more than 17.5 million data points housed in the AAVSO International Database!

Anyone can be a variable star observer. All you really need to begin observing are:

- your unaided eyes, a pair of binoculars, or a telescope
- some variable star charts to help you navigate your way through the sky
- some basic instructions
- a little patience

For those interested in observing activity on our closest star, the sun, or a particular type of variable, such as the Eclipsing Binary and RR Lyrae type stars, or if hunting for novae, supernovae, or optical counterparts to energetic Gamma-Ray Bursts strikes your fancy, we have observing programs designed to help satisfy your appetite.

The AAVSO Mentor Program is available to all observers to assist newcomers in the methods and techniques of visual variable star observation, as well as CCD and PEP observation.

Msgr. Ron Royer, AAVSO member observer since 1953

Services to Astronomy

Mario Motta, M.D., an AAVSO member-observer since 1985, at his 32-inch telescope

The AAVSO provides a wide range of services to the astronomical community. AAVSO International Database data are disseminated extensively to astronomers around the world, upon request, and are freely available from the AAVSO website. AAVSO data and services have been used, referenced, and acknowledged in hundreds of professional astronomical publications.

Services to Astronomers

AAVSO services are sought by astronomers for the following purposes:

- real-time, up-to-date information on unusual stellar activity
- scheduling of variable star observing programs coordinating earth-based large telescopes and instruments aboard satellites
- simultaneous optical observations of program stars and immediate notification of their activity during earth-based or satellite observing programs
- correlation of AAVSO optical data with spectroscopic, photometric, and polarimetric multi-wavelength data
- collaborative statistical analysis of stellar behavior using long-term AAVSO data

Collaboration between the AAVSO and professional astronomers for real-time information or simultaneous optical observations has enabled the successful execution of hundreds of observing programs using satellites such as:

- Hubble Space Telescope
- Chandra X-Ray Observatory
- Spitzer Space Telescope
- XMM-Newton X-Ray Observatory
- Extreme Ultraviolet Explorer
- High Energy Astronomical Observatories 1 and 2
- International Ultraviolet Explorer
- Roentgen Satellite
- European X-Ray Observatory Satellite
- High Precision Parallax Collecting Satellite (HIPPARCOS)

AAVSO services have been used by researchers affiliated with such satellites as Chandra, XXM, RXTE, FUSE, HST, Spitzer, and many more

A significant number of rare events have been observed with these satellites as a result of timely notification by the AAVSO.

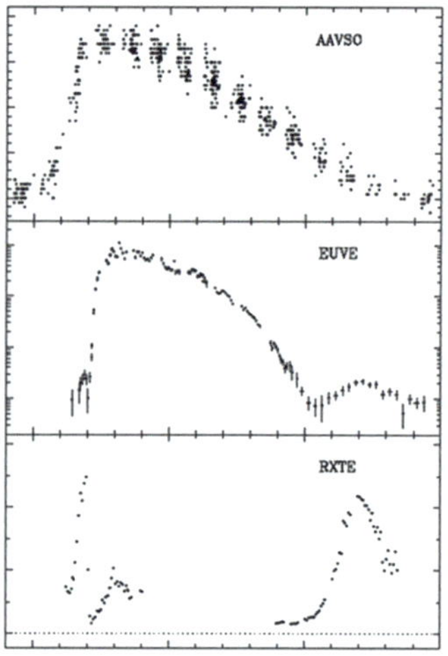

With the outburst detected by AAVSO Observers, simultaneous AAVSO visual, EUVE, and RXTE observations of SS Cygni were triggered, providing astronomers with important information about the behavior of dwarf novae (from Wheatley et al. 2003, MNRAS, 345, 49)

In recent years, the SWIFT satellite has been sending real-time notification to ground- based observers in the AAVSO High-Energy Network to alert them of Gamma-Ray Bursts (GRBs). Several GRB optical afterglows have been detected by AAVSO observers. In this way, AAVSO observers are contributing to cutting-edge, high-energy astrophysics.

Services to Observers and Members

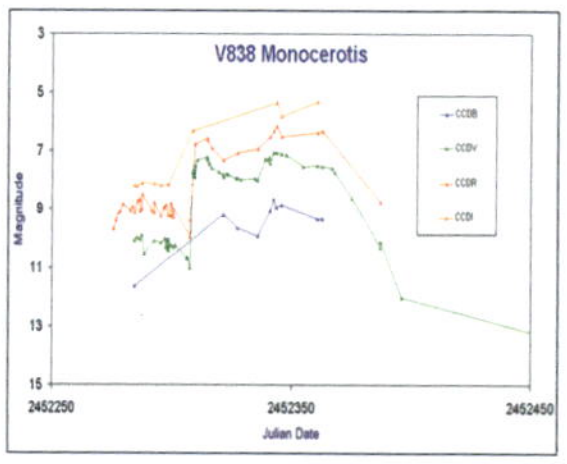

The AAVSO enables variable star observers to contribute vitally to variable star astronomy by accepting their observations, incorporating them into the AAVSO International Database, publishing them, and making them available to the professional astronomer. Incorporating an observer's observations into the AAVSO archives means that future researchers will have access to those observations, so the observer is contributing to the science of the future as well as the present.

The AAVSO coordinates observing campaigns between professional and amateur astronomers, in which observations from amateur astronomers play an important role in correlating observations obtained with special instruments at earth-based observatories or aboard satellites.

On request, the AAVSO will help set up an appropriate observing program for an individual, an astronomy club, an elementary school, a high school, college, and so forth. In this way observers, students, and faculty are able to make the best use of their resources to do valuable science. The AAVSO can also assist in teaching observing techniques and in suggesting stars to be included in a program through the AAVSO Mentor Program.

Gary Poyner, AAVSO member-observer since 1991, with his 14-inch telescope

Education and Outreach

The AAVSO believes that Education and Outreach is important to our mission:

- to attract, train, and retain new variable star observers and members of all ages
- to increase awareness, understanding, and appreciation of variable star astronomy and variable star observing among amateur and professional astronomers, educators, students, and the general public
- to improve science education and literacy through the unique power of variable stars and variable star observing to motivate students, young and old.

Projects, Programs, and Activities

The AAVSO Writers Bureau offers variable star and topical astronomy content on a monthly basis to editors of astronomy club and society newsletters. This gives us the chance to inform the public about the fascinating objects we study, as well as the science and research being done, while providing reliable, accurate information to newsletter editors who may lack the time or expertise to write or vet articles for publication.

The AAVSO has much experience in hosting successful educational lectures such as the series of High-Energy Astrophysics Workshops for Amateur Astronomers

The AAVSO Mentor Program connects experienced observers with new observers to assist them in observing, recording, and reporting observations of variable stars to the AAVSO International Database.

The Speakers Bureau is a service established for people and groups looking for enthusiastic, knowledgeable speakers to provide informative presentations for astronomy clubs, star parties, banquets, Scout Troops, Astronomy Day activities, and other public and private astronomy functions.

Our Presentation Library contains POWERPOINT™ presentations on variable stars, observing techniques and other astronomical topics. These are available free to the to use in making your own presentations.

Variable Star Astronomy (VSA) is an AAVSO educational project, originally developed as Hands-On Astrophysics (HOA) with funds from the National Science Foundation. It is a flexible set of hands-on educational materials, activities, and investigations, based on the AAVSO's unique electronic database of variable star measurements. Students will be able to experience the excitement of doing real science with real data! By carrying out all aspects of the research process, they can develop and integrate skills in science, math, computing, and other areas. VSA has been converted to a web-based format and is expected to be available again in 2009.

VStar is the software that accompanies the activities for VSA. The original DOS-based programs have been ported to a Java platform and are being developed as part of the Citizen Sky project, with funding from the National Science Foundation.

The Education and Outreach Committee

The committee consists of amateur and professional astronomers with a wide range of interests in informal and formal education at all levels, and in many contexts. The committee members are:

Dr. Pamela L. Gay (chair), Southern Illinois University, Physics Department, Edwardsville, Illinois;
Barry B. Beaman, Rockford, Illinois;
Jaime R. García, Instituto Copérnico, Mendoza, Argentina;
Mary A. Kadooka, University of Hawaii, Institute for Astronomy;
Dr. Roger S. Kolman, Harper College, Palatine, Illinois;
Douglas Lombardi, Las Vegas, Nevada;
Paul Mortfield, Thornhill, Ontario, Canada;
Mario E. Motta, M.D., Gloucester, Massachusetts;
Dr. John R. Percy, University of Toronto, Department of Astronomy;
Dr. Pebble L. Richwine Johnson, Riverwatch Middle School, Cumming, Georgia;
Dr. Christine Anne Royce, Shippensburg University, Department of Teacher Education, Shippensburg, Pennsylvania;
Michael A. Simonsen, Imlay City, Michigan;
Arif Solmaz, Çanakkale Onsekiz Mart Üniversitesi, Çanakkale, Turkey;
Donn R. Starkey, Auburn, Indiana;
Donna L. Young, Tufts University, Wright Center for Science Education, Medford, Massachusetts.

2. The Year in Review

Minutes of the 98th Spring Meeting of the AAVSO, Held May 19–21, 2009, Big Bear Lake, California

Gary Walker, Secretary

The 98th Spring Meeting of the AAVSO was held in Big Bear Lake, California, May 19 –21, 2009, as part of a joint meeting of the AAVSO and the Society for Astronomical Sciences (SAS), which held its 28th Annual Symposium on Telescope Science.

The AAVSO Council met Monday, May 18.

Nine paper sessions were held Tuesday morning, May 19, through Thursday afternoon, May 21. Thirty-six papers and posters on various aspects of variable stars, astrometry, photometry, spectroscopy, and occultations were given (see accompanying list for papers presented). On Tuesday evening AAVSO President Paula Szkody and Director Arne Henden led a workshop on Data Mining. The following evening Jerry Foot and Brian Warner led a workshop on Photometry Essentials.

The AAVSO membership meeting was called to order by President Paula Szkody at 3:30 p.m. on Thursday, May 21.

Secretary Gary Walker read the minutes of the 2008 Annual Meeting held in Nantucket, Massachusetts.

Treasurer David Hurdis presented the Treasurer's Report. He reported a total income for the first half of the year of $620,612 and total expenses of $537,165. He also reported that the Endowment fund had decreased from $17.4 M to $9.8 M as a result of the financial crisis in the stock market.

The Director gave his semiannual report. The highlights included an update on AAVSOnet robotic telescope network. Arne reported that we had serviced forty monitoring requests and eighteen calibration requests through the network over the past six months. He reported that the Wright28 telescope is operational at the Krajci

Observatory (renamed Astrokolholz) and that several time series requests have been placed there. Many of these requests were from professionals and first-time amateurs. The availability of telescope time has resulted in collaboration on several NASA/NSF grants. James Bryan has donated an Optomechanics 10-c spectrograph which will be installed on the Morgan Telescope. The AAVSO database now totals approximately 17 million observations. These include the archival AAVSO eclipsing binary and RR Lyr star databases previously maintained by Marvin Baldwin and Gerry Samolyk. In addition, the Henden Calibration database, with 600,000 USNO stars and 400,000 SRO observations, is available as a single MySQL database. He also reported that the AAVSO photoelectric photometry (PEP) software to reduce observer data was rewritten in Perl to accomodate the MySQL database, and the PEP web pages were rewritten. Arne also reported that the AAVSO was awarded a National Science Foundation grant for the International Year of Astronomy (IYA) 2009 in the amount of $400,000 over three years.

Director Henden also reported on the loss of several members and friends: William Albrecht, Patrick Collins, Hugh Maddocks, Paolo Maffei, James Molnar, and Paul Wright. Everyone rose for a moment of silence in their memory.

Reports from several of the new AAVSO Sections (replacing Committees) were presented. Mike Simonsen reported on the Cataclysmic Variable and Transient Section; Section Leaders are Mike Simonsen and Gary Poyner. The Science Advisors are Paula Szkody and Boris Gansicke. The website is operational, as is the watch list.

Dave Hurdis reported on the Short Period Pulsator Section. Gerry Samolyk wants to chair this Section. It was suggested that Dave and Gerry find a co-chair to help out.

Kate Hutton reported on the Long Period Variable Section. She and Mike Simonsen are the Section Leaders. Science Advisors are John Percy, Laszlo Kiss, Lee Ann Willson, and Matthew Templeton.

Activities of the Education and Outreach Section were described by Section Leader Pamela Gay and by Mike Simonsen. They reported issues with IYA support, in terms of the professional community being swamped by IYA activities. The Speakers Bureau and Writers Bureau now have a presence on the web page. The Mentor Program reported that forty-six students were mentored with a 50-50 split between visual and CCD observers.

Arne reported that the IT/Database Section will be organized by Michael Koppelman.

Mike Simonsen gave an update on the progress of the Sequences Section. The Variable Star Plotter (VSP) is working well; a scale charts are now available in addition to the other scales. The Automated Chart Plotter is operational but lots of labor has been required behind the scenes. The sources for the chart data are the 1-m USNO and SRO Henden data. The sequence plotting program SeqPlot shows the field-of-view from SRO and USNO.

After the Membership meeting, the Northwoods Resort Conference Center was converted to a Banquet Hall for the Dinner and Awards Ceremony. Recipients of the AAVSO Director's Award were announced: Sebastian Otero of Buenos Aires, Argentina, and Patrick Wils of Hever, Belgium. Observer Awards were annouced and were presented to those observers in attendance. Following the awards, Dr. John Percy's talk, on "Galileo's Legacy," connecting Galileo's discoveries with modern astronomy and IYA 2009, was enjoyed by all.

The meeting was popular with both AAVSO and SAS members. The synergy between groups was evident. It was suggested that the AAVSO pursue holding every Spring meeting at Big Bear. The meeting was economical, as this time slot is just before the season opens, so the room rates are very low. In addition, the Riverside Telescope Makers Conference is held following the SAS meeting, and would be another enticement to attend. RTMC was attended by several of the AAVSO faithful, including Arne Henden, Kate Hutton, Richard Miles, Barry and Carol Beaman, and Gary Walker.

Attendees at the AAVSO's 98th Spring Meeting held at Big Bear Lake, California

Papers and posters presented at the Joint Meeting of the Society for Astronomical Sciences and the American Association of Variable Star Observers, Held in Big Bear Lake, California, May 19–21, 2009

Arne A. Henden	"The AAVSO Wide-Field Photometric Survey"
Kate Hutton and Michael Simonsen	"AAVSO Long Period Variable Section Update"
Lee Snyder	"BL Eri: A Contact Binary System"
John Menke	"The Addictive Properties of Occultations"
Scott Degenhardt	"High Resolution Asteroid Profile by Multi-Chord Occultation Observations"
Russell M. Genet	"Lightweight Mirror Developments"
Tom Krajci	"Optimizing Opto-mechanical Performance Using Simple Tools and Techniques"
Wayne Watson	"Enhancements to the Sentinel Fireball Network Video Software"
Jay M. Pasachoff	"Photometry and Light Curves in the Solar System"
Richard Miles	"Sloan-r′ Photometry of Comet 17P/Holmes Beyond 3.8 AU: An Observing Methodology for Short-period Comets Far From Perhelion"
Thomas G. Kaye and David Healy	"Spectrashift Exoplanet Transit Search Project: 40,000 Light Curves and Counting"
Daniel O'Connor	"ILOX-A Small Visible Imager on the Lunar Surface"
James Edwards	"Thinking Out Loud: An Optical SETI Campaign Suitable for Amateur Astronomers?"
James W. Young and Alan W. Harris	"The Early History of Photometric Observations of Asteroids Made at Table Mountain Observatory"
Gary A. Vander Haagen	"What's Next in Asteroid Photometry?"
Brian D. Warner	"Slow Rotating Asteroids: A Long Day's Journey Into Night"
Robert Stephens and Ralph Megna	"Extending a Spectroscopic Survey of Main Belt Asteroids With Micro Telescopes: A Proof of Concept Project"
Pamela Gay and Michael Simonsen	"Filling Your Astronomy Program"

Olivier Thizy	"Spectroscopic Binaries Studies"
Nicholas J. Wilsey and Matthew M. Beaky	"Revisiting the O'Connell Effect in Eclipsing Binary Systems"
Rev. Paul Temple	"Using a Web Cam CCD to Do *V*-Band Photometry"
Donald F. Collins and Anesh Prasai	"Intrinsic Variability of β Lyrae Observed With a Digital SLR Camera"
David Boyd and Boris Gaensicke	"An Intensive CCD Photometry Campaign to Observe DW Ursae Majoris"
Jerry D. Horne	"New Observations of Three Lyra Variables"
Robert E. Stencel and Jeffrey L. Hopkins	"ε Aurigae, 2009: The Eclipse Begins—Observing Campaign Status"
Jeffrey L. Hopkins and Robert E. Stencel	"ε Aurigae Hydrogen-a Emission Line Variation: The Horn Dance"
John Pye, Lauren Elder, and Jeffrey Hopkins	"The 2009 Eclipse of EE Cephei: An Educational and Collaborative Journey"
Robert K. Buchheim	"The Light Curve of UZ Sagittae"
Kurt A. Fisher	"An Estimate of the Integrated Magnitude of the LCROSS Impact Ejecta Dust Curtain for Exposure Calibration Practice"
Erin M. Craine	"Data Mining Techniques Applied to the GNAT Library Archive" (poster)
Bandon Decker and Matthew M. Beaky	"Phase-Dependent Spectroscopic Monitoring of Cepheid Variable Stars" (poster)
Amanda Tougas and Matthew M. Beaky	"Searching for Chaos in the Mira Variable Star U Cygni" (poster)
Robert La Pointe, E. Hill, L. Leimer, K. McMillian, A. Miller, and A. Prindle	"Time Delay Integration: A Wide-Field Survey Technique" (poster)
John R. Percy, Samantha Esteves, Alfred Lin, Christopher Menezes, and Sophia Wu	"Quantifying 'Irregularity' in Pulsating Red Giants" (poster)
Lee Snyder	"The Over-Contact Binary GR Tauri" (poster)
Jeff Horne	"Photometry of Variable Stars Using a Lensless Schmidt Camera" (poster)

New Members Accepted at the Spring Meeting, May 19, 2009

Allen, Douglas, IA
Anderson, David, England
S Benson, Steven, GA
Bornak, Jillian, NM
Boyd, David, England
Braga, Raffaello, Italy
Brindle, Peter, England
Brockway, Jack, VA
Bruhn, Carl, Denmark
Bryden, Donald, CA
Burns, Christophe, WA
Campbell, Thomas, MN
Connors, Martin, CT
Costello, Marlin, CA
Crary, Lawrence, FL
Culver, Roger, CO
DeCoster, Rich, IL
Dingle, Larry, CA
Doyle, Courtney, WY
Finch, David, MA
Flateau, Davin, OH
Flechsig, Gerd-Uwe, Germany
Ford, Mike, KS
Fowler, Douglas, OH
Friedman, Vanessa, AZ
Fuqua, Stephen, CA
Galli, Gianni, Italy
Gamble, Geoffrey, MT
Garrett, Russ, AL
Genet, Russell, CA
Goderya, Shaukat, TX
Graham, John L., OH
Grendler, Jean, OR
Hartmann, Kevin, MA
Hearst, Kevin, CA
Hilliard, Wayne, VT
James, Christophe, England
Jenner, Simon, England
Johnson, Carol, CA
Keys, David, FL
Knight, Carl, England
Ludington, Whit, NC
S Majors, David, CA
Maravelias, Grigorios, Greece
Martignoni, Massimilia, Italy
Merhebi, Bob, Lebanon
Miller, Ian, England
Montague, Fred, MA
Moore, John, England
Muyllaert, Eddy, Belgium
Oltion, Raymond, WY
Padovan, Stefano, Germany
Palazzi, Luigi, Italy
Patten, James, NJ
Peairs, Thomas, VT
Perkins, William, MN
Prentice, John, CO
Rachlin, John, MA
Ratcliffe, Martin, KS
Rea, Robert, New Zealand
Reeves, Terry, TN
Rogge, Ralph, Germany
Ross, Hal, FL
Sadauskas, Jonas, Australia
Sawyer, George, UT
Scott, Ryan, Canada
Semien, Antoine, LA
Sewchok, Michael, PA
Shears, Jeremy, England
S Sitko, Michael, OH
Spampinato, Joseph, PA
Steffey, Phillip, FL
Stritof, Niko, Slovenia
Swain, Robert, FL
Sykes, Mark, OH
Temple, Paul, NM

continued on next page

New Members Accepted at the Spring Meeting, May 19, 2009, continued

Tieman, Brian, IL
Veliz, Claudio, VT
Vrenios, Alex, AZ
Walthall, John, TX
Ward, Glen, WV
Weiss, William, CA
Wells, William, OK
West, Robin, Australia
Westall, John, NC
Wilbur, Janet, LA
Williams, Donald, MD
Williams, Richard, CA
Woerner, Edwin, KS
Young, David, Australia
Zimmermann, Thomas, Germany

S = *sustaining membership*

Deceased Members, Observers, and Colleagues

Albrecht, William B., HI
Collins, Patrick D., CT
Maddocks, Hugh C., VA
Maffei, Paolo, Italy
Molnar, James, Jr., VA
Wright, Paul F., MN

AAVSO Director's Award Recipients

Sebastian Otero of Buenos Aires, Argentina was awarded the AAVSO Director's Award "... in recognition of his unique ability to achieve high precision visual estimates and his mentoring of others to help them reach their visual observing potential, and of his contributions to reviewing submissions to the International Variable Star Index (VSX) and guiding observers through the submission process."

Sebastian Otero

Patrick Wils of Hever, Belgium was awarded the AAVSO Director's Award "... in recognition of his contributions to the field of data mining, and of his volunteer effort in support of the International Variable Star Index (VSX), including importing many published lists of stars, reviewing submissions, and acting as a mentor to observers who have discovered new variables."

Patrick Wils

AAVSO Observer Awards

Presented at the 98th Spring Meeting, Big Bear Lake, California, May 19–21, 2009

*Over 150,000 Visual Observations**

Gary Poyner	England	1991–2008	158,493

*Over 100,000 Visual Observations**

Edward A. Halbach	USA	1934–2005	100,016

Presented to him on the occasion of his 100th birthday, April 4, 2009

*Over 25,000 Visual Observations**

Pavol A. Dubovsky	Slovakia	1999–2008	44,201
Alan Plummer	Australia	2001–2008	28,263
Eric Morillon*	France	1988–2008	27,228
Hubert Hautecler^	Belgium	1999–2008	26,356
Thomas Lazuka	USA	1983–2008	25,316

*Over 10,000 Visual Observations**

Tomas Gomez=	Spain	1985–2008	12,417
Vyacheslav M. Ivanov$	Russia	1998–2008	11,921
Erhan Eker	Turkey	2000–2008	10,808
Peter Maurer@	Germany	1993–2008	10,276
Tim Parson	USA	2000–2008	10,141

*Over 350,000 CCD/PEP Observations**

Christopher T. Middleton%	South Africa	2004–2008	375,784	CCD

*Over 200,000 CCD/PEP Observations**

Robert A. James	USA	1953–2008	201,365	CCD

*Over150,000 CCD/PEP Observations**

Tonny Vanmunster^	Belgium	1976–2008	150,686	CCD

*Over 100,000 CCD/PEP Observations**

Bart Staels^	Belgium	1994–2008	129,035	CCD
Giorgio Di Scala	Australia	2004–2008	101,276	CCD

continued on next page

Observer Awards, cont.

*Over 50,000 CCD/PEP Observations**

Ray E. Tomlin	USA	2006–2008	79,734	CCD
David Boyd[<]	England	2003–2008	68,586	CCD
Steve Brady	USA	2004–2008	63,440	CCD

*Over 25,000 CCD/PEP Observations**

Jerry Bialozynski	USA	2004–2008	48,320	CCD
Jeremy Shears[<]	England	2004–2008	29,497	CCD

*Over 10,000 CCD/PEP Observations**

Rudy Poklar	USA	2001–2008	17,559	CCD
James M. Roe	USA	1972–2008	16,719	CCD
Etienne Morelle[*]	France	2005–2008	16,501	CCD
Andrezej Arminski	Poland	2002–2008	13,310	CCD
Carlo Gualdoni	Italy	1983–2008	11,391	CCD
Teofilo Arranz	Spain	2005–2008	10,223	CCD

*Over 2,500 PEP Observations**

Nik Stoikidis	Greece	1974–2008	2,517	PEP

* Years include total AAVSO observing interval (not only PEP/CCD observing).
Total includes only visual or PEP/CCD observations, depending on award.

These symbols indicate observers are also affiliated with the groups below:

- * Association Française des Observateurs d'Étoiles Variables (AFOEV)
- $ Association of Variable Star Observers "Pleione" (Russia)
- % Astronomical Society of Southern Africa, Variable Star Section
- < British Astronomical Association, Variable Star Section
- @ Bundesdeutsche Arbeitsgemeinschaft für Veränderliche Sterne e.V. (BAV) (Germany)
- = Madrid Astronomical Association M1 (Spain)
- ^ Vereniging Voor Sterrenkunde, Werkgroep Veranderlijke Sterren (Belgium)

Minutes of the 98th Annual Meeting of the AAVSO, Held November 5–7, 2009, Newton, Massachusetts

Gary Walker, Secretary

The 98th Annual Meeting of the AAVSO was held at the Crowne Plaza Hotel in Newton, Massachusetts, November 5–7, 2009.

The AAVSO Council met at AAVSO Headquarters all day, Thursday, November 5. They then joined those meeting attendees who had been registering at the hotel during the afternoon, and the meeting began unofficially with an informal dinner at UNO's Pizzaria near the hotel. Being reacquainted with old friends and making new ones was a splendid finish to a long day.

The Council held a Futures Session at AAVSO Headquarters on Friday morning, November 6. Future five-year directions of the organization were discussed, with details to be made available as the plans are executed. The Futures Meeting adjourned at noon.

On Friday morning, a Special Session was held on The Citizen Sky Project and You, by Rebecca Turner, Aaron Price, and Michael Umbricht, describing the AAVSO and partners' major International Year of Astronomy 2009 project and inviting attendees to participate in the many aspects and activities of the project. More information on Citizen Sky is available at www.CitizenSky.org.

After lunch, two workshops were held: Photometrica, by Geir Klingenberg and Michael Kran, on the data reduction program Photometrica, followed by Advanced CCD Observing Techniques, by Arne Henden. Details are available on the AAVSO web site.

Many of the attendees participated in the AAVSO Headquarters Open House and Dinner Buffet on Thursday evening. Headquarters buzzed with all the conversations and reunions taking place!

The Membership Meeting was called to order by President Paula Szkody on Friday morning at 9:02 a.m. Secretary Gary Walker read a shortened version of the Minutes from the 98th Spring Meeting in Big Bear, California.

Treasurer Gary Billings read the Treasurer's report. He reported expenses for FY 2009 of $1,281,330, as opposed to a budget of $1,037,950. Next year's budget is $1,129,522, including all grants. The endowment is currently valued at $12.3M, which is a recovery from the minimum of $9.8M. Current withdraw for FY 2010 from the Endowment will be $717,000, as opposed to $839,000 for last year.

Short Period Pulsators (SPP) Section Leaders Gerry Samolyk and Dave Hurdis reported that Howard Smith and Matthew Templeton have agreed to be Science Advisors.

In the Eclipsing Binaries Section, the legacy of 200 eclipsing binary stars will continue. Section Leader Gary Billings reported recruiting Ed Guinan as science advisor. Dirk Terrell, who has contributed much time and wisdom over the years in eclipsing binary-related matters, has indicated that he must cut back on support for this group. A vote of thanks was given Dirk for all his help.

Long Period Variable (LPV) Section organizer Kate Hutton reported that its legacy list of 100 stars and 390 other important stars was assembled. The creation of a web site is complete. The LPV section is now focusing on which stars to observe, based on science needs.

Education and Outreach Section Leader Pamela Gay reported that IYA 2009 has impacted this section. Also, through his position as President of the Massachusetts Medical Society, and with the cooperation of the American Medical Association, the AAVSO's Mario Motta, M.D., has pushed for adoption of light pollution by-laws nationwide.

Mike Simonsen reported that the Speakers and Writers Bureaus presentation library totals fourteen items so far. He reported thirty-one requests for mentors, and fifty-seven mentored observers this period. Thanks were expressed to Mike Simonsen for all his efforts in these areas.

IT Databases/Data Mining Section Leader Mike Koppelman reported goals of education training, support, and aid to existing programs with data mining tasks. Doug Welch has agreed to be the Science Advisor. A Google discussion group has been set up, with thirty participants to date.

The goals of the Cataclysmic Variable and Transients Sections are to provide information as to what is active and direction as to what should be observed based on new science needs. The most current information on activity should be available at a glance.

Speaking for the Charts Section, Mike Simonsen reported that lots of behind-the-scenes work has been underway on the AAVSO Chart Plotter ACP now called the Variable Star Plotter (VSP). The CHart Error Tracking tool CHET is now online.

First Vice President Jaime Garcia gave a presentation on the Spring 2010 meeting to be held in Valle Grande, Mendoza, Argentina, April 15–18, 2010. The closest large city is San Rafael. Meeting events include sharing a star party with 120 to 180 attendees under spectacular Southern skies, and a trip to the Pierre Auger Cosmic Ray Observatory. Expected costs will be $1,275 air fare, $85 per day for double occupancy, including meals.

Director Arne Henden reported the passing this past year of AAVSO members and friends William Albrecht, Julias Cahn, Patrick Collins, Hugh Maddocks, Paolo Maffei, Thomas Bruce Tregaskis, James Molnar, Hugh Rumball-Petre, Richard Wend, and Paul Wright. Everyone stood to honor them with a moment of silence.

Arne Henden then gave the Membership report and his Annual Director's report. This year we have added 143 new members, 106 from the United States and 37 from abroad. Highlights from the Annual Report include: The AAVSO International Database totals approximately 18 million observations, with 1.2 million new observations this year. Publications included four issues of the *Journal of the AAVSO*, *AAVSO Bulletin 72*, 19 *Alert Notices*, 44 *Special Notices*, 3 issues of the *AAVSO Newsletter*, 56 papers with AAVSO co-authors, and 45 papers in ADS with AAVSO cited as a reference. The AAVSO's telescope network AVSOnet has five telescopes operational, with two more expected before the end of this year and two more next year, for a total of nine. There is also the possibility of more adjunct systems, fully funded through grants and donations. Doug George donated copies of Maxim DL and Bob Denny donated copies of ACP Scheduler to AAVSOnet. Use of AAVSOnet is a membership benefit and will be available to all members in good standing. Sonoita Research Observatory (SRO) calibrated 422,000 stars this year. Two VO interrupts for observations of GRB090418A and GRB091024 were accommodated. Forty proposals have been received for use of SRO for Pro-Am Collaboration. The Wright28 (C11/ST-7xme) is operational. The Wright30 (Meade 12-inch and ST8XME) is operational. BSM (the Bright Star Monitor) is operational; BSM is a 1.5 x 2.2 degree FOV, Takahashi FS-60CB f/6.2 telescope with BVRI filters, and a ST-8XME CCD Camera. APASS (the AAVSO Photometric All-Sky Survey) uses twin ASA astrographs, on a Paramount on loan, located at Tom Smith's Dark Ridge Observatory. The Electronics and Optics lab is operational and is able to characterize CCD cameras. Steve Levine of USNO has joined AAVSO for a year as part of his sabbatical and has set up the lab, written software, and is helping with testing. Regarding the AAVSO International Database, removal of the validation file and its replacement for starname lookup through the International Variable Star Index (VSX) is underway. A field was added to identify the source of observations submitted to the AAVSO International Database from organizations or from individuals who wanted their data identified with a specific organization. AAVSO Webmaster Kate Davis went to Drupal school (Drupal is a leading web design tool). Margaret Mayall Assistant Hannah DiCicco spent the summer summer finishing the scanning of the Eggen cards, which were then placed on line by volunteer Stuart Goldman. The Santa Barbara Instrument Group (SBIG) has been very generous in loaning cameras for various AAVSOnet and other projects. The IYA Citizen Sky and VStar grants were received. Many observing campaigns were organized and executed, and several are continuing. Solar Observer Awards were announced (see accompanying list). In summary, despite the economic situation, it was a good year, and although the economic environment requires some belt tightening, next year should be a very good one.

The membership meeting was adjourned at 10:25 a.m.

Three Scientific Paper Sessions were held, one in the morning and two in the afternoon (see accompanying list of papers presented).

The hall was cleared at 5:30 p.m. after the last scientific talk to allow the room to be reset for the evening.

A cash bar and informal reception were held starting at 6:30, followed by a buffet style banquet which received rave reviews from everyone. Following dinner, numerous awards were announced and presented; the citations for most of these awards follow this report and are also available on the AAVSO website. The 42nd AAVSO Merit Award was awarded to Lewis M. Cook, who, unfortunately, was unable to make the long trip from California to the meeting. The William Tyler Olcott Distinguished Service Award went to Karen J. Meech, and as she could not be present, it was later sent to Jim Bedient for presentation to Karen in Honolulu. Honorary AAVSO membership was bestowed on a very surprised Charles E. Scovil, who was present to receive his standing ovation. A Special Recognition Award was presented to Caroline Moore, who, at 14, is the youngest person to discover a supernova (SN 2008ha). She is a very active amateur astronomer, strongly supported by her parents, who were in the audience. Headquarters volunteer Arthur Ritchie was thanked for his ten years of service with a Special Recognition Award which was presented to him later at a celebration at Headquarters. Staff service awards were presented to several indivuals: Matthew Templeton (7 years); Katherine Davis, Gloria Ortiz Cruz, and Aaron Price (10 years); Gamze Menali (11 years), Michael Saladyga (25 years), and Elizabeth Waagen (30 years).

The Olcott Tankard

Following the awards, Elizabeth Waagen presented the Association with a special gift in recognition of her thirty years with the AAVSO. A large silver engraved tankard that had belonged to William Tyler Olcott was placed for sale last year by its owner, and Elizabeth wanted to see it at "home" with another of Olcott's precious belongings, the AAVSO. She purchased it and has donated it to the AAVSO in thanks for the thirty years of joy the Association has given her. An article about the tankard appears in *AAVSO Newsletter 43*.

Next on the evening's schedule was the keynote speaker, Dr. Steve Howell. His engrossing talk was titled The NASA Kepler Mission: Planets for Everyone and had everyone anticipating after he predicted that several major announcements would be made at the AAS meeting in January, and after he hinted that there may be follow-up work on exoplanets which are not part of the Kepler Mission (i.e., they are not believed to be Earth-sized planets in the habitable zone).

As the final event of the meeting and her final duty as President, outgoing President Paula Szkody handed the gavel over to incoming President Jaime Garcia. Paula was soundly thanked by everyone for her dedicated efforts over the past two years. President Garcia then thanked everyone for coming, said he looked forward to seeing them in his native Argentina in April, and closed the meeting at about 11 p.m.

Papers and posters presented at the Scientific Paper Session on Saturday, November 7th, 2009

Edward J. Los	"Estimate of the Limiting Magnitudes of the Harvard College Observatory Plate Collection"
John Pazmino	"The Park in the Sky"
Mary Ann Kadooka Mimi Hang	"Hawaii Student/Teacher Astronomy Research (HI STAR) Outcomes"
Gary Billings	"Rapid Cadence Monitoring of ε Aurigae"
Steve B. Howell	"Kepler Observations of Variable Stars"
Barbara L. Welther	"Mrs. Fleming's "Q" Stars."
Paula Szkody	"GALEX and Optical Light Curves of LARPS"
Donald F. Collins	"Intrinsic Variability of Eclipsing Variable β Lyrae Measured With Digital SLR Camera"
Mike Simonsen	"The Z CamPaign"
Michael Koppelman	"Making Good Plots With Excel"
Kristine Larsen	"Scientists Look at 2012: Carrying on Margaret Mayall's Legacy of Debunking Pseudoscience
Grant Foster	"T UMi: from Mira to ???"
Dave Hurdis Tom Krajci	"Secular Variation of the Mode Amplitude-Ratio of the Double-Mode RR Lyrae Star NSVS 5222076"
Andy Howell	"*BVRI* Photometry of W UMa Binary Systems and Lessons Learned"
John Percy Samantha Esteves Jou Glasheen Alfred Lin Marina Mashintsova Sophia Wu	"Variability 'Profiles' for T Tauri Variables and Related Objects, From AAVSO Visual Observations"

New Members Accepted at the Annual Meeting, November 5, 2009

Beckstrom, Garry, MI
Bennett, Christophe, FL
Birriel, Jennifer, KY
Blanco Gon, Jesus, Spain
Bologna, Robert, MI
Bonnardeau, Michel, France
Brooks, Mike, CO
Cason, Andrew, GA
Davis, Clay, NM
Dose, Eric, TX
Dudley, Robert, PA
Durkin, Michael, NY
Engstrom, William, PA
Erickson, Mark, CA
Euman, Charles, IL
Finley, Otis, CA
Gedam, Subhash, CT
Goldberger, Martin, NY
Gwyn, David, PA
Hambsch, Franz-Jose, Belgium
S Hannon, Kevin, MD
Hardies, Mark, FL
Holahan, Gary, MD
S Hoofnagle, Dennis, WA
Howard, John, AL
Johnson, Donald, TX
Kalajian, Peter, ME
King Salz, Elizabeth, HI
Kucharski, Thomas, CA
Ladowski, Theodore, WI
Lake, Peter, Australia
Lanagan, Peter, NV
Lopez, Jorge, TX
MacLeod, Morgan, ME
Manuel, Jose Anton, Spain
Martin, John, IL
McKown, Archibald, MD
Meriaux, Jean-Chris, CA
Ngeow, Chow Choon, IL
O'Connor, Linda, MA
Olsen, Aart, IL
Pearson, Robert, VA
Sakamoto, Takanori, MD
S Smith, Thomas, MD
Sokolovsky, Kirill, Germany
Starr, Peter, Australia
Stegeman, James, MO
Thompson, Mark, CA
Truax, Jonathan, MI
Williams, Bobby, GA
Williams, Peter, Australia
Wood, William, AZ

***S** = sustaining membership*

Deceased Members, Observers, and Colleagues

Cahn, Julius, NM
Rumball-Petrie, Hugh, CA
Tregaskis, Bruce, New Zealand
Wend, Richard, IL

AAVSO Merit Award Recipient (announced at the 98th Annual Meeting in Newton, MA, November 7, 2009)

Lewis M. Cook was awarded the 42nd AAVSO Merit Award "... in recognition of over 35 years of dedicated service to the AAVSO as variable star observer, contributing nearly 160,000 observations, including timely observations of campaign stars; Council member and officer for 11 years; developer of observing tools and software, and a leader in the development of CCD observing standards and practices; and for his generous spirit and tireless enthusiasm for observing variable stars which have earned him the respect and admiration of astronomers around the world."

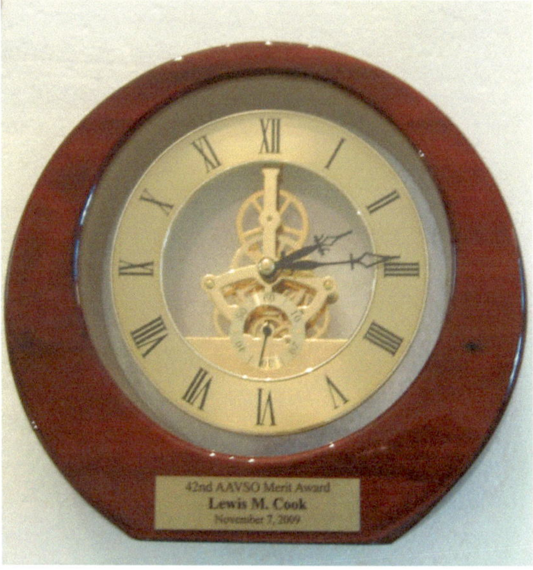

The AAVSO Merit Award clock to be presented to Lew Cook

AAVSO William Tyler Olcott Distinguished Service Award Recipient (announced at the 98th Annual Meeting in Newton, MA, November 7, 2009)

Karen Meech was awarded the AAVSO William Tyler Olcott Distinguished Service Award "... for her promotion of variable star observing through her distinguished work as astronomer and educator, her service to the AAVSO as Council member and colleague, and her being both friend and inspirational role model to teachers and students of astronomy."

Special Recognition Award Recipients (presented at the 98th Annual Meeting in Newton, MA, November 7, 2009)

A Special Award of Recognition and honorary AAVSO membership was presented to ***Charles E. Scovil***, "In recognition of his devotion to science, public education, and the ideals of the Association, and for the manner in which he has represented the AAVSO for over fifty years as councilor, officer, committee chair, *Atlas* and chart preparer and advisor, *Journal* associate editor, *Circular* co-editor, and contributor of over 37,000 visual observations to the AAVSO International Database."

Charles Scovil

Caroline Moore

A Special Award of Recognition was presented to ***Caroline Moore*** "...in recognition of her accomplishment as the youngest person to discover a supernova, and her contribution to science in that Supernova 2008ha in UGC 12682 was an unusual, potentially rare, type of supernova event. The AAVSO admires her patience and perseverance and considers her a role model for teens, girls, women, data miners, and citizen scientists everywhere."

Special Recognition Award Recipient (announced at the 98th Annual Meeting in Newton, MA, November 7, 2009)

Arthur Ritchie received a Special Award of Appreciation "...for ten years of devoted volunteer service to the Association through assisting the Solar Committee Chair in digitizing Sunspot reports and helping with all manner of tasks at Headquarters, large and small, exciting and mundane, with cheerful enthusiasm and efficiency, sharing his kindness of spirit and wisdom with his colleagues."

AAVSO Solar Observer Awards (presented at the 98th Annual Meeting in Nantucket, MA, November 7, 2009

Sunspot Observers (1,500 or more observations)

Patrick Abbott, Canada
Howard Barnes, New Zealand

Sudden Ionospheric Disturbance Observers (40 or more months of reports)

Roberto Battaiola, Italy
Andy Clerkin, Massachusetts
Michael Hill, Massachusetts
Michael King, England
Lionel Loudet, France
Jon Wallace, Connecticut

Annual Report of the Director for Fiscal Year 2008-2009

Arne A. Henden, Director

We've had a remarkably busy year, even with the poor economy. AAVSOnet was extended, private funding was found for many projects, and close to another two million observations were submitted to theAAVSO International Database. The big news was the award of a major NSF grant to the AAVSO. I'll open with that story!

The AAVSO and the International Year of Astronomy 2009

The International Astronomical Union (IAU) voted in 2006 to create IYA2009, the International Year of Astronomy. The intent was to publicize astronomy and to bring it home to the masses. The IAU did not provide any funds for this project; instead, each member country was to develop their own national plan and both fund it and provide a means for grass-roots participation.

Throughout 2007, the AAVSO (in the form of Aaron Price) was a key member of the Citizen Science working group for the U.S. national office of IYA2009. In November, the AAS submitted a National Science Foundation (NSF) grant to fund the national program managers, as well as four selected theme projects. One of those projects was Citizen Science, using the rarely eclipsing variable star epsilon Aurigae as the star of the project. NSF granted the proposal, but severely cut the funding, basically supporting the national office but not any of the four theme projects.

Aaron Price then largely wrote an NSF Informal Science Education (ISE) grant proposal, with major partners the AAVSO, Adler Planetarium, the California Academies of Science, and the University of Denver, that was submitted in June 2008. In December 2008, we received word that our proposal was under consideration for award, but that a number of questions had been raised by the review panel. We worked for the next nine months to answer those questions and to demonstrate our ability to handle this large grant. In September 2009, we were finally awarded the three-year grant. Now, some details about the project!

citizensky

The Citizen Sky Project

Citizen Sky (www.citizensky.org) is the AAVSO's citizen science project for the International Year of Astronomy (IYA 2009). It began on September 1, 2009, when the National Science Foundation awarded the AAVSO a three-year grant to fund its development and operation. Currently there are over 2,000 project participants from over twenty countries.

Who are citizen scientists? Volunteers, many of whom have no prior scientific training, who work with trained scientific researchers to answer real-world questions.

A photograph of epsilon Aurigae taken by Alson Wong with a Vixen102-ED telescope and a STL-11000M CCD camera.

The "Star" of the Project, epsilon Aurigae (eps Aur). This is a bright variable located in the constellation Auriga, the Charioteer. At third magnitude the star is bright enough to be seen with the unaided eye even in the most light polluted cities. It is well placed for observing in the fall, winter, and spring skies. Every 27.1 years the star goes into an eclipse that lasts approximately one and a half years. Even after over 175 years of scrutiny this variable is not fully understood. The current eclipse of eps Aur began in August of 2009, will continue throughout 2010, and will end in early 2011. Eps Aur is particularly well suited to the Citizen Sky project because:

- It is bright enough to be observed without specialized equipment making it easy for beginners to get involved;
- The beginning of the current eclipse happened to coincide with the International Year of Astronomy 2009;
- Even though this object is bright and has been studied over a long period of time there are still many important questions to be answered.

The AAVSO's Citizen Sky project welcomes anyone who has an interest in participating regardless of previous experience. In fact, no previous experience at all is required. We are guiding new participants through the process of how to observe eps Aur, how to send observations of the star to the AAVSO, how to see their results, analyze them, and even publish them at the end of the project. Many citizen science projects stop at data collection and hand the data off to "real" scientists for analysis and publishing. We hope that this will be the largest citizen science project in modern history that takes its participants through the entire scientific process resulting in real, active research.

Along with press releases and various interviews about the project, we have developed tools to recruit and educate participants about epsilon Aur and Citizen Sky. Here are a few of them:

- Website: A completely new website was created for the project by AAVSO Web Developer, Kate Davis, Citizen Sky Project Manager, Rebecca Turner, and other Citizen Sky staff. This site serves as the "home base" for anyone interested in participating in the project. The Citizen Sky website includes blogs, discussion forums, topical chats, polls, an events calendar, a place to submit and analyze data, training materials, plus much more.

- Workshops: A three-day workshop, focused on observing and education/public outreach, was held in August 2009 at the Adler Planetarium in Chicago. There were over fifty participants in this workshop from over twenty states spread across the U.S. Participants were selected based on their ability and willingness to return home after the workshop and disseminate the information they acquired. Video of talks from the August workshop are available on the Citizen Sky website. A second workshop on data analysis and scientific paper writing will be held at the California Academy of Sciences in September 2010. Workshop updates and application instructions will be posted on the Citizen Sky website in early 2010.

- Visualizations: Several participants have put both their artistic skills and astronomy knowledge to good use and developed artistic renderings of certain models of the epsilon Aurigae system. Two professional visualizations are also near completion. A planetarium show trailer will be available for download by Spring 2010. An interactive visualization is also near completion and will allow users to adjust parameters such as star mass and disk radius.

Illustration of the top view of the epsilon Aurigae system based on the most popular model at this time. Artwork by Citizen Sky participant Nico Camargo.

Participation in the Citizen Sky Project can be broken down into five stages. We hope that as many of our participants as possible will proceed through all of the stages, allowing them to participate in the full scientific process.

1. Getting Started: epsilon Aurigae is an ideal target for those interested in learning how to observe variable stars. The Citizen Sky website is home to loads of background information and tutorials—including the Ten Star Tutorial. This tutorial trains new observers in the techniques needed to make and report visual variable star estimates. Training materials are available on topics such as: observing, analysis, publishing, and outreach.

2. Submitting and Viewing Data: Participants are able to use the Citizen Sky website to easily submit the data that they have collected to the AAVSO International Database. Data points can be viewed instantly as a table or light curve.

3. Staying Motivated and Connected: Once participants submit ten observations, they receive a Citizen Sky certificate to thank them for their contribution and to encourage them to keep observing. There are many ways for participants to stay connected including: project blogs, forums on various topics, and a monthly Citizen Sky newsletter. Participants may also form or join Citizen Sky teams if they wish to collaborate with others. Teams are made up of individuals with different yet complimentary skill sets who wish to work together toward a common goal (publishing a paper, developing software, etc.).

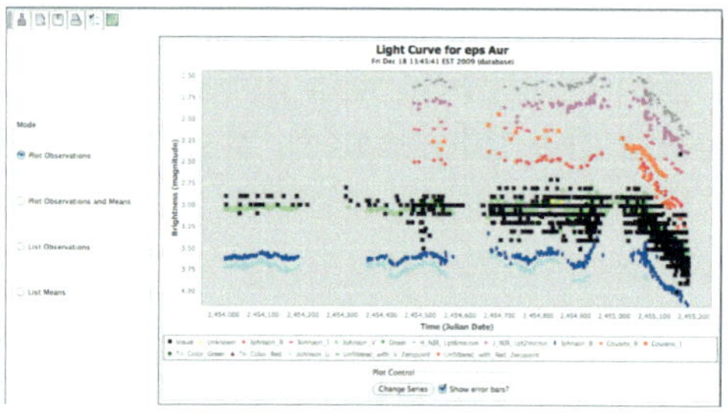

Screen shot from VStar.

4. Analyzing Data: Data analysis tutorials will be online in early 2010. There is a Citizen Sky forum focusing exclusively on data analysis. Also, a new data analysis software package called VStar has been developed by Citizen Sky participants and is in the beta testing stage. The Citizen Sky workshop being held in September 2010 will focus on data analysis techniques.

5. Publishing Results: The September 2010 workshop will also include instruction on scientific paper writing. A special edition of the *Journal of the AAVSO* will be dedicated to papers written by Citizen Sky participants.

The next couple of years will be quite informative as we continue to watch this mysterious eclipse unfold. We are excited about the new information that will come both from the multiple professional campaigns and our own Citizen Sky results. After 175 years perhaps the mystery of epsilon Aurigae will finally be solved.

Observation Database

In FY2009, we collected approximately 1.84 (1.7) million observations. 788,825 (873,411) of these were visual observations; 13,329 (1,450) were PEP or photographic observations. The remainder 1,034,870 (837,310) were CCD observations. The CCD totals remain high, as we get many thousands of observations for any time-series campaign (SS Cyg is an example). The two charts on the following pages show the annual submission totals since 1911, and the total submitted observations ("Megasteps") since 1911. You can see that the trend is exponential, so that by 2011, we will be collecting 15 million observations per year!

Work continues on importing the RASNZ database. A large fraction of the observational data in it comes from just a few observers, such as Albert Jones and Danie Overbeek, and so was straightforward to import. The remaining observations require assigning

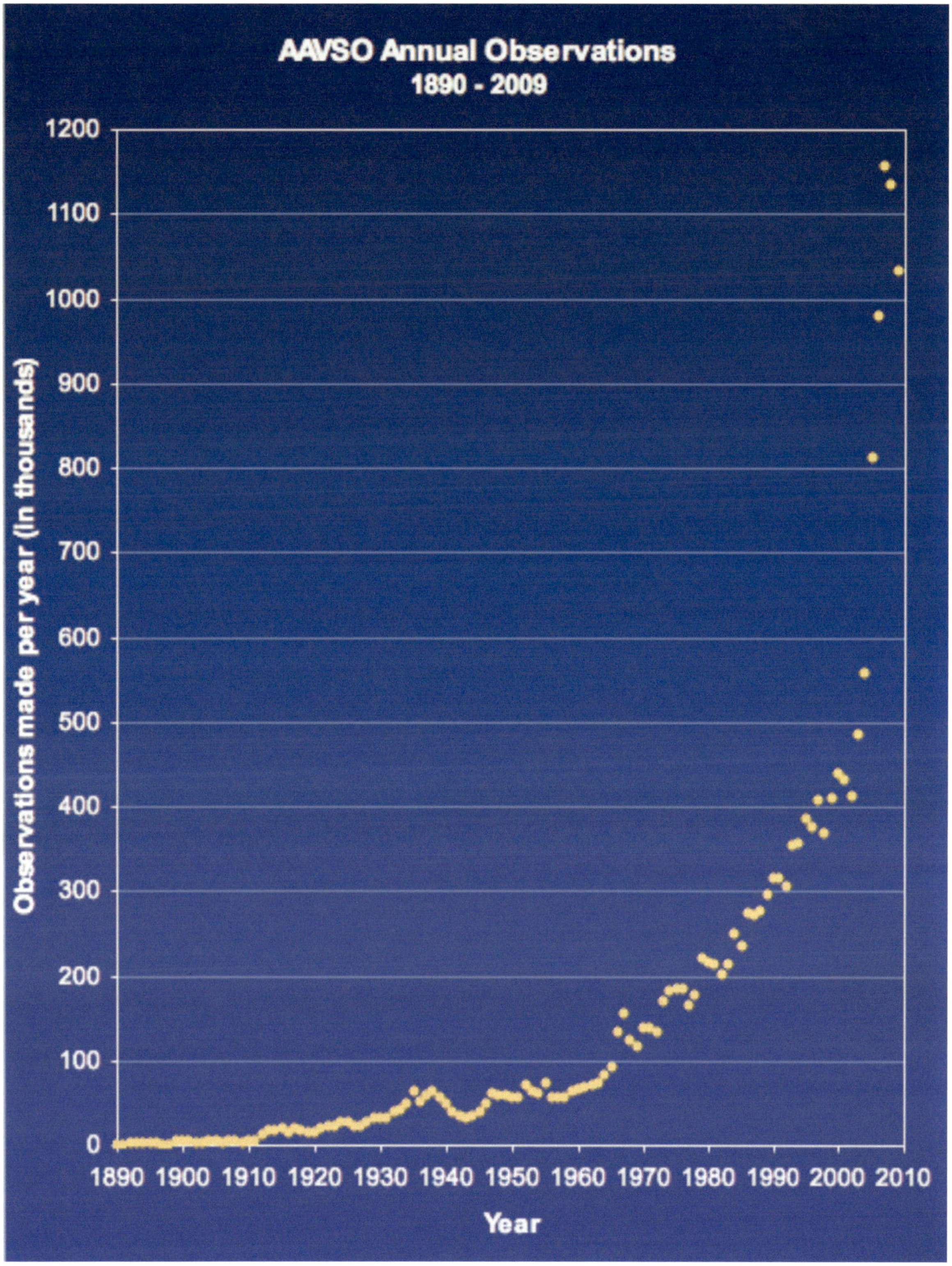
AAVSO Annual Observations
1890 - 2009
Observations made per year (in thousands)
0
100
200
300
400
500
600
700
800
900
1000
1100
1200
1890 1900 1910 1920 1930 1940 1950 1960 1970 1980 1990 2000 2010
Year

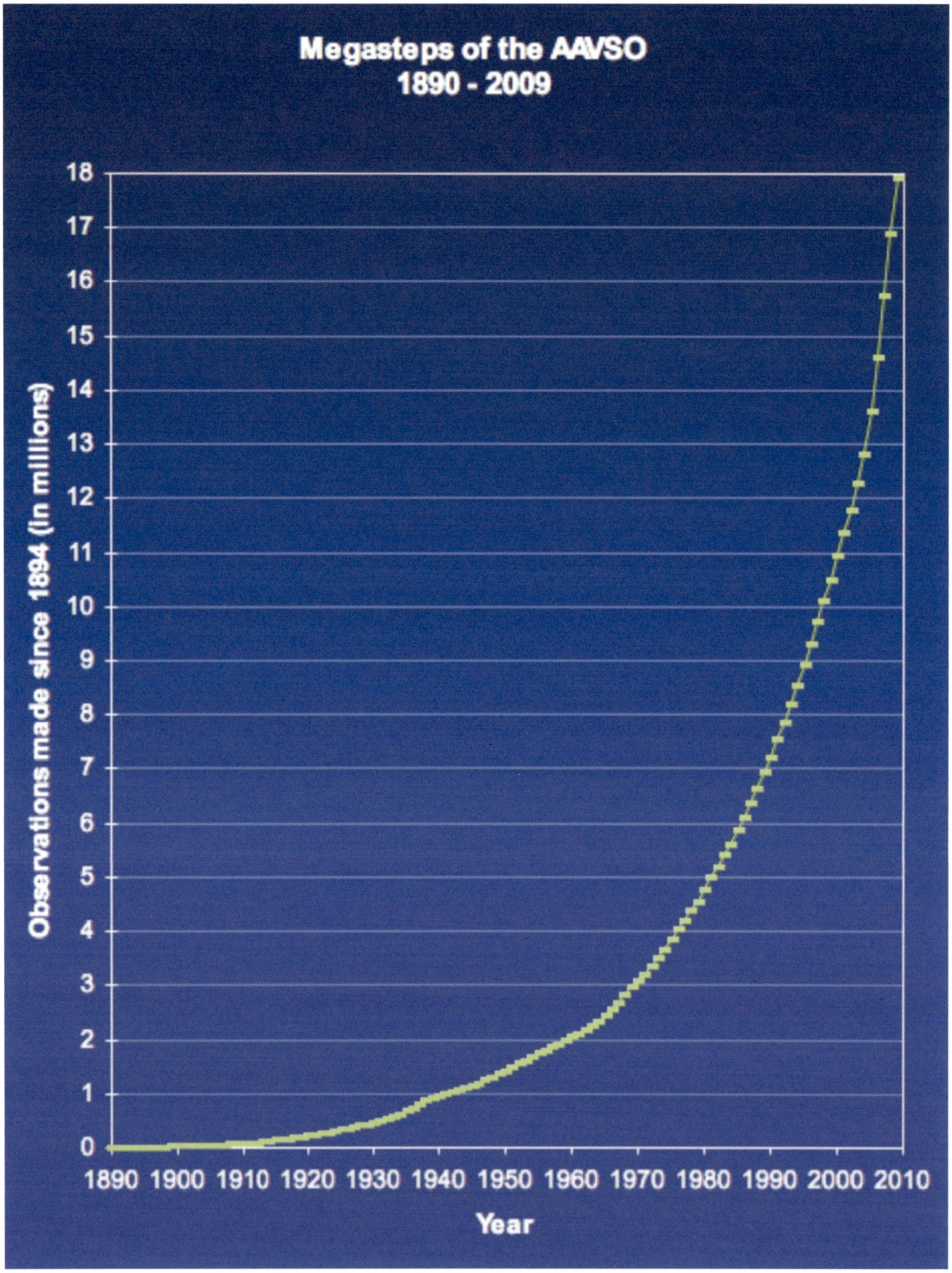
Megasteps of the AAVSO
1890 - 2009
Observations made since 1894 (in millions)
0
1
2
3
4
5
6
7
8
9
10
11
12
13
14
15
16
17
18
1890 1900 1910 1920 1930 1940 1950 1960 1970 1980 1990 2000 2010
Year

observer codes to those many observers who were not AAVSO contributors, as well as determining what charts and comparison stars were used. We hope to finish this project in the near future.

We had 5,242 data requests from a multitude of researchers during the year (see charts on the following pages). The data request rate is pretty constant throughout the year, but has definitely continued its upward trend.

A long-standing problem with the existing MySQL International Database was the creation of duplicate records due to the twin-table schema of the original implementation. Aaron merged these two tables during the summer; all internal programs then had to be updated to use the single-table format.

We've finally added into the International Database all of the eclipsing binary observations that were originally archived by the eclipsing binary chair, rather than HQ. To the best of our knowledge, all observations that have been contributed to the AAVSO now reside in the International Database. You should check your observing totals, and compare them with your local logbooks to make sure that we have everything that you have submitted!

Back in the flat-ascii-text database days, we used a "validation file" that listed all of the stars for which observations existed in the International Database. As observers submitted observations of new stars, a staff member would edit the validation file to add the new star. As we moved into the relational database age, the primary key in the database was the AAVSO Unique ID, or AUID. This increased the complexity of the validation file, as it now had to include the AUID, all possible aliases of the star name, some notes such as the star's R.A. and Dec. so that it could be uniquely identified in other

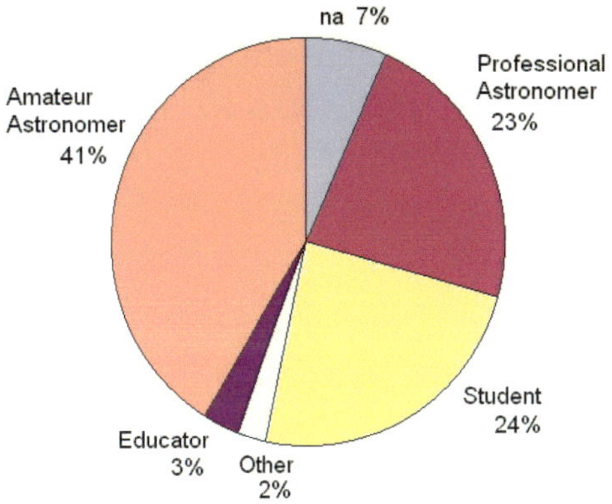

Researchers who used AAVSO data or services during FY 2008–2009

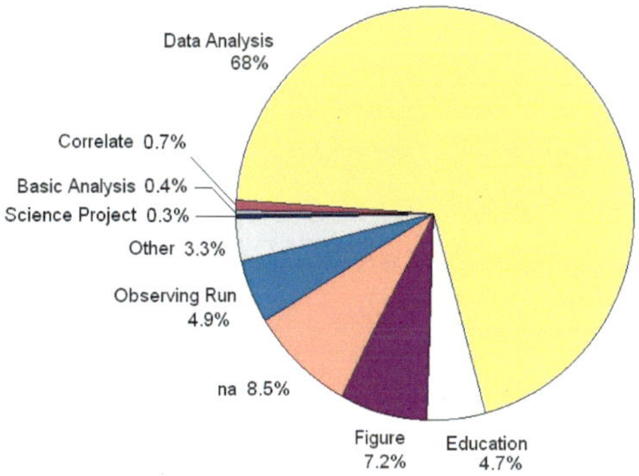

Areas in which AAVSO data or services were used during FY 2008–2009

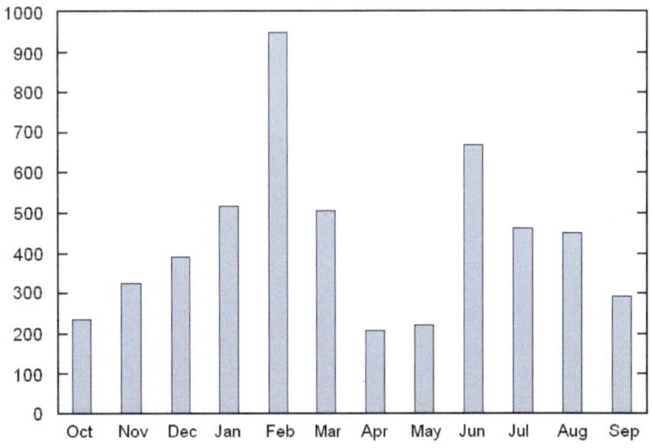

Number of data requests by monh during FY 2008–2009

catalogs, etc. We also wanted to move to a method where observers could submit data on any known variable, as listed in the International Variable Star Index (VSX). This year we converted all of our main tasks to use VSX as the name resolver, rather than the validation file. This is working superbly, and so we have now removed the validation file table from all AAVSO software.

International Cooperation

We acknowledge with appreciation the observations sent to the AAVSO by members of the following variable star associations, either individually or as a group, for inclusion in the AAVSO International Database for dissemination to the astronomical community worldwide:

a. Agrupacion Astronomica de Sabadell (Spain)
b. Asociacion Amigos de la Astronomia (Argentina)
c. Asociacion de Variabilistas de Espagne (Spain)
d. Association of Variable Star Observers "Pleione" (Russia)
e. Association Française des Observateurs d'Étoiles Variables (AFOEV) (France)
f. Astronomical Society of Southern Africa, Variable Star Section
g. Astronomisk Selskab (Scandinavia)
h. Astronomischer Jugendclub (Austria)
i. Brazilian Observational Network REA
j. British Astronomical Association (BAA), Variable Star Section
k. Bundesdeutsche Arbeitsgemeinschaft für Veränderliche Sterne e.V. (BAV) (Germany)
l. Grupo Astronomico Silos (Spain)
m. Israeli Astronomical Association, Variable Star Section
n. Koninklijke Nederlandse Vereniging Voor Weer-en Sterrenkunde, Werkgroep Veranderlijke Sterren (Netherlands)
o. Liga Ibero-Americana de Astronomia (South America)
p. Madrid Astronomical Association M1 (Spain)
q. Magyar Csillagàszati Egyesület, Valtózcsillag Szakcsoport (Hungary)
r. Norwegian Astronomical Society, Variable Star Section
s. Red de Observadores (Montevideo, Uruguay)
t. Royal Astronomical Society of Canada
u. Royal Astronomical Society of New Zealand, Variable Star Section
v. Ukraine Astronomical Group, Variable Star Section
w. Unione Astrofili Italiani (Italy)
x. URSA Astronomical Association, Variable Star Section (Finland)
y. Variable Star Observers League in Japan
z. Vereniging Voor Sterrenkunde, Werkgroep Veranderlijke Sterren (Belgium)

Nikolai Samus (Sternberg Observatory) visited AAVSO Headquarters from April 12 through April 19, staying in the Feibelman Guest Suite. Kolya was here to discuss mutual projects related to the General Catalogue of Variable Stars.

Elizabeth Griffen (U. Victoria) visited the AAVSO in the spring to help define a spectroscopic project.

Software

As an aid to the Chart Team, I had Sara Beck port my sequence-plotting program to Java. Called Seqplot, this program accesses a MySQL database of all of the photometric mean magnitudes and colors that I have determined at Sonoita and USNO. It plots the stars with dot size based on magnitude and color based on the star's (B–V) color. With a cursor you can point to any given star, click, and see its magnitude and colors; with another click, you can move the information into a file in a pre-defined format that can be easily uploaded into the Variable Star Database (VSD), the database used to generate comparison star sequences for charts. Seqplot is now being used by many volunteers world-wide to improve sequences and create new ones, especially those for transient objects in a timely manner.

Kate modified an administrative tool written by Chris Watson for the VSD that enables staff editing and updating of this important database. The program will edit single records as well as accept batch uploads of new sequences, especially those coming from Seqplot.

I had used a really neat photometry software package called Photometrica while helping the Global Rent-a-Scope network improve their scientific imaging capability. The author of Photometrica, Geir Klingenberg, is well known to AAVSO observers, as he wrote WinWWZ. Geir and Michael Kran are working on making Photometrica even better, and they contacted me to see if I would be interested in helping out (testing the photometric algorithms). They have also offered to give full control of the program to the AAVSO for use by our observers. I'll have more to say about this project in next year's report.

Observing News

Ed Halbach reached two important milestones this year. In March, Ed turned 100. While I was not able to attend the party, I understand a good time was had by everyone.

Second, Elizabeth checked the database records and confirmed that Ed had submitted over 100,000 lifetime observations. Gerry Samolyk presented Ed with his Observer's Award at the birthday party.

We had nearly two dozen active campaigns during FY2009. Some of these were just for fun, such as the one to monitor QX Pup (an obscured Mira variable in which the optical variation is a reflection of the star from the bipolar nebula that surrounds it), or the one-in-a-million chance to detect an eclipse of the white dwarf V1412 Aql. This latter star was studied extensively by Arlo Landolt (now on the AAVSO council), after he could not locate the star one night a few decades ago. The assumption is that the star was in eclipse, and since the companion could not be detected, it must be nearly planetary in size and mass. A monitoring campaign was initiated to see if we could detect another eclipse. Michel Bonnardeau created a simulation that uses the available non-eclipse-detection photometry to narrow the possible period choices, and to predict optimal windows in which to search for another eclipse.

Other campaigns were in support of HST cataclysmic variable projects, looking for outbursts of blazars to alert observers using VERITAS, monitoring P Cygni in collaboration with a group of German spectroscopic observers to see if there are correlations between continuum variations and spectroscopic line changes, checking for a possible exoplanet transit for HD 80606, and obtaining precision photometry of beta Cephei in support of interferometric observations. Of course, we're running a campaign on epsilon Aurigae as well!

Members volunteer their time and effort in promoting the AAVSO. Kate Hutton used the AAVSO traveling display at the Riverside Telescope makers Conference in May 2009, with many visitors passing by the booth and picking up our brochure. She and I did the same thing at the new Pacific Astronomy and Telescope Show, held at the Pasadena Convention Center in September 2009. About 1,000 people attended and passed by our booth. Mike Simonsen took the traveling display to ALCON 2009, where he represented the AAVSO along with Gordon Meyers. These meetings are attended by many of the high-end amateurs who may not have considered doing photometry with their equipment, so we are really happy when someone offers to stand at a booth or poster for hours and answer questions.

Several stars decided to do something spectacular in honor of FY2009. R CrB, the prototypical dust-fading star, entered its most recent fade at the beginning of the fiscal year. It has stayed faint the entire year (unusual) and at a *V* magnitude of 15 (also unusual). Since the last fade was over five years ago (and it was a puny fade), this exciting event has made R CrB a favorite target for many observers. When will R CrB regain its normal brightness? GK Per had a small outburst; VY Scl went into a deep fade (and we ran a campaign in support of a VLT program); U Sco stayed in quiescence, even though Brad Shaefer was hoping otherwise; and lots of novae went into outburst. A deep eclipsing cataclysmic variable (CV) was discovered (CSS 081231:071126+440405) and was studied by several observers. Eclipsing CVs are a lot of fun, as the eclipses are extremely short and usually quite deep. You can be taking a time series in which one exposure shows the CV and in the next exposure the star is completely gone! V630 Cas had an unusual outburst, nearly Gaussian in shape, about 2.5magnitudes in amplitude, and lasting 100 days. V630 Cas is well-monitored by a number of CCD observers, with a very nice light curve even though it is a 16th magnitude object. Several groups are now starting to monitor M31, reporting novae in that galaxy. Since these can also be 16th magnitude, it is becoming commonplace to obtain light curves of such extragalactic objects with amateur equipment.

The RASNZ variable star section has taken new life. Headed by Tom Richards, it is now known as Variable Stars South, and has its own discussion group (AVSON) and web site. They are creating several interesting campaigns. If you are a southern observer, I highly recommend participating in some of their projects. We've been passing interesting southern objects to them to make sure that they are involved in any southern campaign.

Robotic Telescope News

The AAVSOnet robotic telescope network is being slowly expanded at low cost to provide access to scientific-grade systems around the world for our membership and our professional collaborators.

Between the SAS meeting and the Stellar Pulsation meeting, I stayed in the southwestern U.S. and visited Tom Smith and Tom Krajci, along with looking at the 24-inch telescope at New Mexico State University. The Smith/Dark Ridge Observatory visit was to see the progress on the Morgan 24-inch telescope. While the telescope itself is still waiting for refurbishment, Tom Smith has been busy building the roll-off roof building that encloses it. I helped put together the drive system to open and close the roof, and had many fruitful talks with Tom Smith and Tom Krajci on their hosting of AAVSO telescopes. I was also able to visit New Mexico Skies and other telescope farms in the immediate area.

We received CCD cameras through a grant from the Santa Barbara Instrument Group (SBIG). An STL-6303 is designated for the SRO50, a 50-cm replacement for the Sonoita Research Observatory (SRO); an ST-8XME is to be used on the Bright Star Monitor, described below; several ST-402's were obtained to be given to observers on long-term basis. We received three ST-6 cameras from a private donor, and distributed them as well as the ST-402 SBIG cameras to new observers. Stephen Levine wrote a Python script to create a pulsed-LED linearity tester in the HQ optics lab, and tested the linearity of all cameras before shipping them to their respective sites.

SRO was used on 244 nights during FY2009 for AAVSO projects, with about half of those being photometric. Many long period variable, Cepheid, and RR Lyr fields were calibrated during this year. A total of 813,361 objects have been calibrated on at least one night at Sonoita using the C14 telescope since 2005. Monitoring of many campaign objects and personal research targets was performed. We slowly released the telescope to the public and initiated observing programs for AAVSO members, with about forty projects currently underway.

James Bryan, an amateur astronomer in Texas, donated an Optomechanics Model 10-C spectrograph to the AAVSO. This is similar to the SBIG DSS-7, though with more features and with replaceable gratings and a gas discharge tube calibration source. We will be modifying this spectrograph for automated use.

Paul Wright, a long-time member of the AAVSO, passed away in November 2008. He donated his two telescopes to the AAVSO. One was already sited at Tom Krajci's Astrokolkhoz Observatory near Cloudcroft, NM. Tom offered to continue supporting that telescope, and to also build an enclosure and support Paul's second telescope. These two telescopes—the W28 and W30—form an essential part of AAVSOnet, providing time-series monitoring of many objects. Gary Walker contributed many filters and a filter wheel for W28; Josch Hambsch contributed an ST-9, filter wheel, and filters for W30. Tom Krajci also heavily modified W30 (an older Meade LX200 12-inch SCT), replacing the metal tube with a carbon fiber one. The AAVSO provided a Meade focal reducer for W30. Patrick Wiggins donated an SSP-3 photometer, which we loaned to Erwin van Ballegoij for work on epsilon Aurigae.

We've started work on an interesting micro observatory. Called the Bright Star Monitor (BSM), it was designed to complement the All-Sky Automated Survey (ASAS), run by Gzregorz Pojmański. BSM will monitor variables in the range V=2 to V=8 at *BVRcIc*, especially concentrating on epsilon Aurigae. We were able to get private funding for the initial BSM hardware, and Tom Krajci has offered to host the system at his observatory.

A key project for the AAVSO's future is the AAVSO Photometric All-Sky Survey (APASS). A grant from the Robert Martin Ayers Sciences Fund was awarded in 2009, and we've purchased nearly all of the necessary hardware for this system. It will calibrate the entire sky from 10th to 17th magnitude in five band-passes. We will install it in Fall 2009 at Tom Smith's Dark Ridge Observatory near Weed, NM. After surveying the northern sky, the system will be moved to the southern hemisphere to complete the sky. We expect APASS to make single-field photometric calibration a thing of the past. It will take about three years to complete the survey, though incremental releases will start just a few months after first light.

Other Projects

As mentioned in the last Director's Report, Donna Young has been bringing Hands-On Astrophysics into the modern media world. That project was completed this year, and we've created a web page for Variable Star Astronomy (VSA). You can download all chapters and teacher exercises for VSA from that page.

The Astronomical League has developed a Variable Star Club. Headed by Rocky Togni, this club will provide finding charts and mentoring for league members so that they can contribute observations to the AAVSO.

The AAVSO speakers bureau and the writers bureau continue to be expanded and improved. The speakers bureau is a list of those people who are willing to give talks on astronomical topics, especially related to variable stars. The writers bureau is a compendium of those bloggers who have given permission for use of their material in club newsletters and other publications. Mike Simonsen is the primary contact for these new initiatives.

The Olin Eggen observation card archive has been completely scanned. Most of the scanning was completed in summer 2007 by David Coit. Linda Henden has worked on scanning the remainder of the cards, along with our 2008 summer student, Sungmun Choi, and our 2009 Margaret Mayall Assistant, Hannah diCicco. Once the scanning was completed, Stuart Goldman, another volunteer, created a web page to publish these cards to the community.

Staffing

Arthur Ritchie continues volunteering at HQ. He comes in whenever we call for assistance, usually to help in stuffing envelopes, mailing solar bulletins, and general sorting. We really appreciate his efforts, and they save considerable staff time. This year marked his tenth anniversary as an AAVSO Headquarters volunteer, a milestone we recognized at the Annual Meeting in November.

Hannah diCicco was our 2009 Margaret Mayall summer assistant, finishing up the Eggen card scanning and beginning the web page for serving these scans to the public. Hanna is *Sky & Telescope* editor Dennis diCicco's daughter and was a joy to have in the office. Since Dennis works across the park at the *Sky & Telescope* offices, father and daughter got together often!

As part of the Citizen Sky project, Rebecca Turner was hired full-time as the Program Manager. Rebecca is well-known to everyone, as she has been our meetings coordinator for a decade, lately performing this important role as a part-time contractor while pursuing a second career in acting.

Virginia Renehan has replaced Travis Searle as our Adminstrative Assistant. Ginny is a past-president of the Amateur Telescope Makers of Boston, and has extensive background in office management.

Dr. Stephen Levine from USNO-Flagstaff is working at the AAVSO this year. While he is primarily telecommuting to Flagstaff, he is lending his computer and hardware expertise where needed on AAVSO projects in his spare time.

Aaron continues his doctoral work in Science Education at Tufts University. He has completed all course work and just has to finish his dissertation.

Other than these changes, headquarters staffing has remained constant. With the new additions, we have twelve full-time employees, along with two part-time employees and a contracted accountant. All permanent employees are described on our website at http://www.aavso.org/aavso/about/staff.shtml. I encourage you to read about these folk that support the members and observers; it is a really nice and efficient staff at HQ!

Publications

Thomas R. Williams and Michael Saladyga continue work on the AAVSO centenary book. They hope to publish by 2011.

As mentioned in the last Annual Report, Charles Whitney retired as Editor in Chief of the *Journal of the AAVSO*. We were able to twist John Percy's arm to "come out of retirement" and take over the position. We published our last hardcopy journal (v35n2); all recent issues are available in pdf format on the web site, or in hardcopy through our Print On Demand (POD) partner. The POD quality is nearly as good as the volume printing, and eliminates a major headache to the staff of envelope stuffing, mailing, and keeping old volumes in stock, and the very significant expense of printing and mailing over a thousand copies of each issue.

As in past years, Tom Williams has funded a "month" in the annual calendar of the American Astronomical Society. We have picked October as our month, and create a new theme every year. The theme for the 2009 calendar was epsilon Aurigae and how the AAVSO was involved in its study. Kerri Malatesta has done an excellent job over the years of designing these calendar pages.

We experimented with creating an AAVSO annual calendar. It contained pictures provided by members and observers, and incorporated the JD calendar. As it was our first one, we were a bit late in production, and that led to disappointing sales. We will skip 2010, but intend to have a special version for 2011.

As part of our policy of continued improvement, we've upgraded the *AAVSO Newsletter*. Edited by Gamze Menali, Elizabeth Waagen, and Michael Saladyga, it now incorporates Eyepiece Views, CCD Views, the Photoelectric Photometry Newsletter, and observing information, as well as the normal articles. It is published quarterly, but only in electronic format. We went that route to save costs, as well as to provide new features such as color and hyperlinks. We also produced the *AAVSO Annual Report* for FY2008.

JAAVSO volume 35, number 2, through volume 37, number 1, were published. Many *eJAAVSO* articles were posted. We posted 19 Alert Notices and 44 Special Notices. Three "Variable Star of the Season" articles were published. We contributed sections for the *RASC Observer's Handbook*. Elizabeth completed long period variable maxima/minima *AAVSO Bulletin 72*. The AAVSO released the annual eclipsing binary/RR Lyrae stars ephemerides as well as the monthly *Solar Bulletin*.

There were fifty-six staff publications (Henden, Price, Templeton, Waagen; *PASP, AJ, JAAVSO*, etc.). We noted that forty-five papers in journals such as *Astronomy and Astrophysics, MNRAS, ApJ, AJ, PASP*, etc. were published using AAVSO data and assistance. The actual number is larger than this, as many posters and papers at AAS meetings use our light curves in their presentations.

Travel and meetings

We held a joint AAVSO and Society for Astronomical Sciences (SAS) meeting in May. SAS is always held just before the annual RTMC meeting, in Big Bear, California. About half of their membership also belong to the AAVSO, so there is good synergy between the groups. The joint meeting was the largest SAS meeting in history, and the paper sessions and workshops were enthusiastically received.

FY2009 was another year of travel by staff to domestic meetings to spread the word about the AAVSO and variable star observing. I would also like to mention that much of my travel is subsidized by the hosts of the meetings I attended. Sometimes they can contribute towards the plane fares, and often provide housing, meals, and logistical support. This is gratefully appreciated!

I went to three specialist's meetings this year. Normally I attend one, but things conspired to bring them all into one year. In October, I went to Huntsville for the Sixth Huntsville Gamma-Ray Burst Symposium 2008, the first GRB meeting since Fermi was launched. I hadn't been to a GRB meeting in about five years, and wanted to see what the latest news was regarding the follow-up activities and where AAVSO observers can contribute. In March, Mike Simonsen and I went to Wild Stars in the Old West II, the 14th North American Workshop on Cataclysmic Variables, in Tucson, hosted by Steve Howell. All of the big names in CV research were there; it was good to talk to some of these folks. Finally, in May, Matt Templeton and I went to Santa Fe to attend the Stellar Pulsation: Challenges for Theory and Observation workshop. This is given every other year, and with CoRoT in orbit and Kepler soon to be launched, there were some very interesting papers. We made some important contacts, and now have representation on the Kepler Asteroseismic Cosmology Consortium.

I went to the June American Astronomical Society meeting in Pasadena, Galileo's Legacy conference in Hawaii over New Year's, and the Pacific Astronomy and Telescope Show (PATS) in September, all where I gave talks regarding the AAVSO. There were two invited trips (Eastern Tennessee State University and the University of Cincinnati) to give advice on near-IR observing and small-telescope projects. Several of us attended the Citizen Sky workshop in Chicago in August to give papers and workshops. Finally, I was invited to sit on a couple of NSF/NASA review panels.

Looking Towards the Future

Coming up over the next fiscal year will be a number of improvements in support of our observers. We will be adding more precision photometry to the comparison star database. APASS will complete its northern hemisphere survey. More campaigns will be announced. The robotic telescope network will be expanded, with both 24-inch telescopes coming on-line. Hopefully some of our submitted grants will be awarded. All-in-all, I think it will be another great year for the AAVSO!

Acknowledgements

This is not a one-person show, or even a dozen-person show. Everyone who has contributed data, made a monetary donation, volunteered their time and energy, has made this organization the success that it is. We "stand on the shoulders of giants" who came before us and built the foundation of the organization. Clint Ford contributed enormously to the organization, which is why his name bears such prominence everywhere. Previous Directors organized the association and had the vision for its future. The Council guides the AAVSO, volunteering their efforts to make the organization financially solvent and relevant. Our committee chairs and section leaders handle specific areas of interest, working with enthusiastic observers and making reports to the membership and Council. Others work quietly behind the scene, acting as scientific advisors to programs, writing important software, or participating in important projects such as the Sequence Team. Finally, many institutions and government agencies see our research important enough to provide financial support. Without all of these people, the AAVSO would not exist.

Observer Totals

Our special appreciation and thanks go to our enthusiastic and dedicated observers, who are the heart of the AAVSO and whose ongoing efforts make this association vital to variable star research. Listed on the following pages are the totals of observations that we received at Headquarters this year and that were made during the year. These totals do not include historical database observations (such as from the RASNZ) or individual historical datasets digitized from the literature.

Table 1. AAVSO Observer Totals 2008–2009 by Country.*

Country	*No. Observers*	*No. Obs.*	*Country*	*No. Observers*	*No. Obs.*	*Country*	*No. Observers*	*No. Obs.*
Argentina	3	47	Germany	38	13744	Portugal	3	4470
Australia	25	81146	Greece	7	5146	Republic of Korea	1	1
Austria	3	649	Hungary	44	20010	Romania	15	6346
Belarus	1	18	India	1	11	Russia	8	1101
Belgium	16	73312	Iran	1	1	Serbia	1	12
Bermuda	1	115	Ireland	4	132	Slovakia	1	1024
Bolivia	3	357	Isle of Man	1	1	Slovenia	1	15
Brazil	14	2092	Israel	1	2	South Africa	6	5250
Bulgaria	4	32	Italy	23	7018	Spain	29	15380
Canada	32	31933	Japan	4	1199	Sweden	1	1774
Chile	4	9380	Lebanon	1	3	Switzerland	5	472
China	3	137	Mexico	1	824	Turkey	2	85
Croatia	4	6868	Netherlands	10	6174	Ukraine	3	129
Czech Republic	2	132	New Zealand	12	5809	Uruguay	2	245
Denmark	6	347	North Cyprus	1	14178	U.S.A.	296	590921
England	25	48707	Norway	7	512	Wales	2	18618
Finland	13	17312	Philippines	2	134			
France	30	95641	Poland	20	8389	TOTAL	743	1097355

Table 2. AAVSO Observer Totals 2008–2009 USA by State or Territory.*

State		*No. Observers*	*No. Obs.*	*State*		*No. Observers*	*No. Obs.*	*State*		*No. Observers*	*No. Obs.*
Alabama	(AL)	1	1	Maine	(ME)	3	7629	Pennsylvania	(PA)	14	2468
Arizona	(AZ)	17	23738	Maryland	(MD)	9	831	Puerto Rico	(PR)	1	17
Arkansas	(AR)	1	42	Massachusetts	(MA)	18	23553	Rhode Island	(RI)	3	3128
Army Post Office	(AE)	1	13	Michigan	(MI)	6	3848	South Carolina	(SC)	2	32
California	(CA)	38	13724	Minnesota	(MN)	7	3429	Tennesee	(TN)	2	121
Colorado	(CO)	5	5784	Mississippi	(MS)	1	691	Texas	(TX)	24	13503
Connecticut	(CT)	4	1093	Missouri	(MO)	2	12683	Utah	(UT)	1	844
Fleet Post Office	(AP)	1	7	Montana	(MT)	1	12404	Vermont	(VT)	2	17
Florida	(FL)	11	110732	Nebraska	(NE)	2	58	Virginia	(VA)	9	149
Georgia	(GA)	6	3398	New Hampshire	(NH)	3	538	Washington	(WA)	9	5281
Hawaii	(HI)	2	589	New Jersey	(NJ)	2	9	West Virginia	(WV)	2	985
Illinois	(IL)	13	155419	New Mexico	(NM)	8	95314	Wisconsin	(WI)	7	54444
Indiana	(IN)	9	2553	New York	(NY)	10	5527	Wyoming	(WY)	2	114
Iowa	(IA)	3	38	North Carolina	(NC)	6	474				
Kansas	(KS)	5	1506	Ohio	(OH)	14	1818	TOTAL		296	590921
Kentucky	(KY)	1	29	Oklahoma	(OK)	3	67				
Louisiana	(LA)	4	196	Oregon	(OR)	3	22083				

** Totals do not include historical data (data preceding fiscal 2008–2009) submitted during 2008–2009.*

Table 3. AAVSO Observers, 2008–2009.*

Code	Org.	Name	No. Obs.
AFO		A. Abascal, Spain	5
AAP	27	P. Abbott, Canada	3892
ACSA		C. Acosta, AZ	5
ARV		R. Adamson, CA	19
ACN	13	C. Adib, Brazil	671
SBX		S. Adrian, Romania	6
ASA		S. Aguirre, Mexico	824
AWL		W. Alexander, VA	12
ASAS3		All Sky Automated Survey 3, Chile	9351
ACO	20	C. Allen, Sweden	1774
ADL		D. Allen, IA	2
AWH	14	W. Allen, New Zealand	1
ARP		R. Allnut, TX	22
AJC	13	J. Almeida, Brazil	108
AJV	15	J. Alonso, Spain	199
AAA	13	A. Alves, Brazil	18
AAQ	03	A. Ambrus, Hungary	1
AAX	13	A. Amorim, Brazil	957
AJE		J. Andrei, Romania	1
CIQ		C. Andronie, Romania	3
AJN		J. Appleyard, Canada	142
AMIB		M. Armstrong, England	1
ACP		P. Armstrong, TX	4656
ARJ		J. Arnold, TX	64
ATE		T. Arranz, Spain	3821
ATO	08	T. Aslesen, Norway	13
ATI	03	T. Asztalos, Hungary	2360
ADI	02	D. Augart, Germany	457
BOZ	03	B. Bago, Hungary	647
BIY		D. Bailey, IL	3
BIE	05	A. Baillien, Belgium	411
BANT		A. Baker, MI	5
BFX		R. Baker, OH	33
BWW		W. Bakewell, CA	1
BFO	03	J. Bakos, Hungary	969
BAH		A. Balcerek, Poland	2
BALJ	14	A. Baldwin, New Zealand	18
BIV	03	I. Balogh, Hungary	16
BGZ		G. Banialis, IL	175
BZV	03	Z. Baracki, Hungary	4
BHAA		H. Barghamadi, Iran	1
BSR	18	S. Baroni, Italy	227
BPO		D. Barrett, France	1745
BQ	03	L. Bartha, Hungary	4295
BVT		T. Bartlett, TX	510
BBA		B. Beaman, IL	2056
BWX	27	A. Beaton, Canada	505
BSZ		S. Beckwith, MA	9
BDQ		A. Bedard, WA	184
BJS		J. Bedient, HI	201
BCP	20	C. Beech, England	1027
BZX		G. Beltran, Bolivia	339
BTY		T. Benner, PA	166
BEB		R. Berg, IN	130
BQX	15	M. Betlej, Poland	34
BIZ		J. Bialozynski, AZ	17165
BVO		V. Bibe, Argentina	11
BQM		M. Bignotti, Italy	1
BCO		C. Birza, Romania	3
BXN	01	M. Bisson, France	248
BXT	08	T. Bjerkgaard, Norway	225
BKL		J. Blackwell, NH	188
BPB		P. Blanchard, CA	128
BVZ		J. Blanco Gonzalez, Spain	36
BLD	10	D. Blane, South Africa	278
BWZ		E. Blown, New Zealand	379
BREI	02	R. Boettcher, Germany	29
BHQ	29	T. Bohlsen, Australia	1792
BPF		P. Bohnholdt, Denmark	27
BBF		B. Boller, VA	3
BVS		S. Bolzoni, Italy	15
BZU		M. Bonnardeau, France	1484
BRJ		J. Bortle, NY	4224
BMU	04	R. Bouma, Netherlands	7
BDG	20	D. Boyd, England	16766
BMK		M. Bradbury, IN	96
BXS		S. Brady, NH	340
BNW	02	W. Braune, Germany	50
BQC	01	J. Breard, France	216
BTB		T. Bretl, MN	264
BHA	02	H. Bretschneider, Germany	873
BMI		M. Brewster, TX	8
BQE		E. Briggs, Canada	42
BHP		K. Brinson, FL	4
BJQ	27	J. Brooks, CA	26
BXV	15	X. Bros Caton, Spain	42
BMB		M. Brown, PA	13
BPR	01	P. Brunet, France	14
BOA	01	A. Bruno, France	7537
BHU		R. Buchheim, CA	103
BXD		A. Burda, Romania	1209
BIW		N. Butterworth, Australia	7105
CCB		C. Calia, CT	248
CMN		R. Cameron, Australia	15
CPN	27	P. Campbell, Canada	21
CMP		R. Campbell, FL	979
CEM	15	E. Capella, Spain	12
CVJ	06	J. Carvajal Martinez, Spain	13
CLQ		L. Cason, SC	30
CWO		W. Castro, OH	61
CQJ		J. Centala, IA	28
CNT		D. Chantiles, CA	453
CGF		G. Chaple, MA	1794
CFX		F. Char, Chile	5
CKJ		J. Cheng, PA	10
CQS		S. Cheng, China	120
CCY		C. Chiselbrook, GA	361
CCX		C. Chrestani, Brazil	4
CCV		C. Clarasso, Spain	39
CLK		W. Clark, MO	1
CPE	06	P. Closas, Spain	57
CDK		D. Collins, NC	432
COL		P. Collins, AZ	4
CME	18	E. Colombo, Italy	260
CTIA		T. Colombo, Italy	47
CMG	04	G. Comello, Netherlands	2835
CDSA		D. Conner, England	9
CKL		A. Cook, OH	56

Table 3. AAVSO Observers, 2008–2009, cont.*

Code	Org.	Name	No. Obs.	Code	Org.	Name	No. Obs.
COO		L. Cook, CA	210	FMZ		M. Fitzgerald, TX	5
CLZ	01	L. Corp, France	203	FLE		L. Florin, Romania	23
CAI		A. Correia, Portugal	2234	FDA	03	A. Fodor, Hungary	71
COV		V. Coulehan, NY	45	FBZ	03	B. Fodor, Hungary	18
CFY		J. Craig, MA	27	FMR		M. Fonovich, Croatia	6836
CTX		T. Crawford, OR	20888	FJQ		J. Foster, CA	1612
CMD	20	M. Crow, England	172	FEX		E. Fox, PA	1
CRR		R. Crumrine, NY	39	FXJ		J. Fox, NM	68
CTI	03	T. Csorgei, Hungary	209	FML	04	M. Fridlund, Netherlands	3
CSM	03	M. Csukas, Romania	373	FCHA		C. Froeschlin, Germany	10
CKB		B. Cudnik, TX	1981	FMG		G. Fugman, NE	51
CUU		J. Curto Amigo, Spain	229	FDX		D. Fuller, IL	2
DAH	08	H. Dahle, Norway	7	FRTA		R. Fuller, TX	16
DHO		H. Dale, GA	1	FSC		S. Fuqua, CA	73
DQA		A. Dandrea, FL	161	GBZ	21	O. Gabzo, Israel	2
DAM	06	A. Darriba Martinez, Spain	114	GHT	27	G. Gaherty, Canada	142
DMP		M. Dasgupta, India	11	GMO		M. Gainer, PA	8
DJX	27	M. De Jong, Canada	198	GJSA		J. Galang, WI	2
DPP	05	P. De Ponthiere, Belgium	20907	GGL	18	G. Galli, Italy	132
SWQ	13	W. De Souza, Brazil	61	GME		J. Gardner, CA	11
DSJ	13	J. De Souza Aguiar, Brazil	8	GAA		P. Garey, IL	67
DKEA		K. Deakes, Isle of Man	1	GBL		B. Gary, AZ	1
DDFA		D. Dempf, Germany	5	GKI		K. Geary, Ireland	27
DDE		D. Denisenko, Russia	1	GCP	02	C. Gerber, Germany	201
DLAA		L. Depka, Germany	23	GHS		H. Gerner, WI	273
DNO		O. Deren, Poland	31	GQR		R. Gherase, Romania	4
DSI		G. Di Scala, Australia	24464	GAO		A. Giambersio, Italy	2
DPA	05	A. Diepvens, Belgium	24	GGU	04	G. Gilein, Netherlands	246
DRG		R. Diethelm, Switzerland	400	GMY		M. Glennon, Ireland	14
DLA		A. Dill, KS	77	GZN		A. Glez-Herrera, Spain	222
DIL		W. Dillon, TX	475	GFT	01	F. Gobet, France	883
GDB	03	G. Domeny, Hungary	21	GFB		B. Goff, CA	6495
DPV		P. Dubovsky, Slovakia	1024	GPU		P. Goldfinger, CA	7
DAB		A. Dukes, SC	2	GOT	06	T. Gomez, Spain	2489
DMO	01	M. Dumont, France	702	GED		E. Goncalves, Brazil	4
DMPA		M. Durkin, NY	19	GVG		V. Gonzalez Garcia, Spain	8
DKS		S. Dvorak, FL	107804	GGZ	03	Z. Gorgei, Hungary	154
DGP		G. Dyck, MA	749	GHN		J. Graham, OH	246
EED		E. Edinho, Brazil	11	GKA		K. Graham, IL	27428
EJF		J. Edmonds, MA	1	GPE		J. Grainger Observatory, NH	10
EHEA		H. Eggenstein, Germany	5	GRL	08	B. Granslo, Norway	49
EMA		M. Eichenberger, Switzerland	12	GDQ		D. Gray, FL	6
ELE		L. Elenin, Russia	13	GNJ		J. Green, Canada	3
EM		G. Emerson, NM	2	GBD		B. Griffiths, New Zealand	16
EPE	01	P. Enskonatus, Germany	36	GTZ		T. Grzybowski, NM	2341
ERB		R. Eramia, WA	48	GCO		C. Gualdoni, Italy	3962
EJO	03	J. Erdei, Hungary	929	GUN	01	J. Gunther, France	287
EEY		E. Erdelyi, CA	126	GSHA		S. Gupta, TX	1
EMJ		M. Erickson, CA	17	GGX	01	G. Guzman, France	171
EFE		F. Etchart, Argentina	3	HCS	03	C. Hadhazi, Hungary	1781
EDTA		D. Etscorn, WY	6	HDH	03	S. Hadhazi, Hungary	258
ERW	14	R. Evans, New Zealand	51	HTY		T. Hager, CT	735
FAZ		A. Falzolgher, Italy	23	HKB		B. Hakes, IL	178
FSU		S. Fanutti, Canada	17	HJW		J. Hall, CO	6
FAM		A. Farkas, AP	7	HYD	14	D. Hambly, New Zealand	8
FEO	03	E. Farkas, Hungary	194	HMB	05	F. Hambsch, Belgium	11613
FAJ	03	A. Fejes, Hungary	4	HJCA		J. Hancock, TX	6
FAF		A. Few, WA	6	HDX		D. Hands, NC	2
FEV		E. Fischler, WA	30	HPL		P. Hansen, Denmark	81

Table 3. AAVSO Observers, 2008–2009, cont.*

Code	Org.	Name	No. Obs.	Code	Org.	Name	No. Obs.
HIC	03	I. Hanyecz, Hungary	6	KAD	03	A. Karpati, Hungary	238
HBB		B. Harris, FL	52	KEI		E. Kato, Australia	8
HMQ		M. Harris, GA	116	KBJ		R. Kaufman, Australia	18
HKM		K. Hartmann, MA	9	KPI	17	P. Kehusmaa, Finland	120
HHU	05	H. Hautecler, Belgium	2196	KSH	29	S. Kerr, Australia	232
HKY	27	K. Hay, Canada	5	KJJ		J. Keski-Jylha, Finland	450
HAB		R. Hays, IL	817	KSZ	03	S. Keszthelyi, Hungary	240
HMH		M. Heald, AE	13	KEA	03	R. Kilmaj, Hungary	12
HPC		P. Hecht, Germany	6	KRB		R. King, MN	564
HRZ	30	R. Hegenbarth, Germany	1	KQR		R. Kinne, MA	5
HRH		R. Hensler, Australia	9	KSJ	27	S. Kinsella, Canada	61
HES		C. Hesseltine, WI	1270	KIR		P. Kirby, AZ	373
HMV		M. Hessom, CA	13	KIL	03	L. Kiss, Australia	190
HJJ		J. Hewlett, CA	12	KMM	09	M. Kititsa, Ukraine	82
HDEA		D. Higgins, NY	4	KPC		P. Klages, England	12
HJX	13	J. Hodar Munoz, Brazil	2	KPL		P. Kneipp, LA	55
HEK	11	E. Hoeg, Denmark	82	KCD	20	C. Knight, New Zealand	2
HFO	01	G. Hoffer, Germany	35	KGT		G. Knight, ME	36
HDF		D. Hohman, NY	51	KSP		S. Knight, ME	70
HSQ		S. Holland, NC	2	KLO		L. Kocsmaros, Serbia	12
HYA	14	A. Homes, New Zealand	58	KRV		R. Koff, CO	2604
HJL		J. Homes, New Zealand	42	KHL		M. Kohl, Switzerland	18
HOO	04	G. Hoogeveen, Netherlands	49	KYI		Y. Kok, Australia	136
HPO		J. Hopkins, AZ	317	KZN	03	Z. Kolarovszki-Sipiczki, Hungary	19
HJG		J. Horne, CA	10	KRS		R. Kolman, IL	1636
HJZ		J. Horne, CA	17	KMA		M. Komorous, Canada	3315
HSR	05	S. Hoste, Belgium	108	KJK		J. Konasek, Czech Republic	125
HSP	14	S. Hovell, New Zealand	611	KMP		M. Koppelman, MN	350
HSW		S. Howerton, KS	178	KOS	03	A. Kosa-Kiss, Romania	4131
HJA		J. Hudson, CA	53	KLX		L. Koscianski, MD	13
HDU		D. Hurdis, RI	3126	KAF	03	A. Kovacs, Hungary	262
HUR	20	G. Hurst, England	2864	KVI	03	I. Kovacs, Hungary	219
HTN		K. Hutton, CA	1443	KWO	02	W. Kriebel, Germany	1719
HUZ		R. Huziak, Canada	5906	KIS	02	G. Krisch, Germany	398
IAT		A. Ielo, Italy	9	KTV	16	T. Kryachko, Russia	78
ILE	03	E. Illes, Hungary	449	KTZ		T. Krzyt, Poland	232
JPM	10	P. Jacobs, South Africa	36	KBA		B. Kubiak, Poland	190
JJB	11	J. Jacobsen, Denmark	38	KUC	01	S. Kuchto, France	2411
JMA		M. Jacquesson, France	186	KPB		P. Kuebler, OH	31
JTP	01	P. Jacquet, France	116	KBO		R. Kuplin, AZ	6
JM		R. James, NM	75401	KAPB		A. Kurtz, MA	1
JZO	03	Z. Jankovics, Hungary	515	KMI	16	M. Kuzmin, Russia	10
JKK	08	K. Jensen, Norway	64	KSQ		S. Kuznetsov, Russia	971
JLR		R. Jepeal, CT	48	LCR	15	C. Labordena, Spain	565
JGE	06	G. Jimenez, Spain	6	LHS		H. Lacombe, Canada	34
JDKA		D. Johnson, TX	4	LMU		M. Lahteenmaki, Finland	8
JOG		G. Johnson, MD	92	LSA	17	S. Lahtinen, Finland	1
JA	14	A. Jones, New Zealand	4546	LPB		P. Lake, Australia	88
JJI		J. Jones, OR	1182	LDJ	27	D. Lane, Canada	1662
JPGA		P. Jordanov, Bulgaria	14	LTO	02	T. Lange, Germany	2
JTDA		T. Judah, CA	17	LMF	13	M. Lara, Brazil	229
JAZ	03	A. Juhasz, Hungary	12	LTM		T. Laskowski, IN	20
JWM		W. Julian, NM	2015	LZT		T. Lazuka, IL	782
KMY		M. Kaczmarech, Brazil	1	LEB	01	R. Lebert, France	143
KJGA		J. Kade, MI	2	LJF	27	J. Lebold, Canada	5
KPK		P. Kalajian, ME	7523	LMT		M. Legutko, Poland	286
KB		W. Kaminski, NM	9	LDA		D. Lehman, MD	2
KMO		M. Kardasis, Greece	253	LDI		D. Lehmann, Germany	3
KSF		S. Karge, Germany	166	LPD	01	P. Lemarchand, France	49

Table 3. AAVSO Observers, 2008–2009, cont.*

Code	Org.	Name	No. Obs.	Code	Org.	Name	No. Obs.
LNZ		G. Lenz, LA	101	MIW	20	I. Miller, Wales	18612
LJL		J. Leonard, IL	12	MLL		J. Miller, MD	2
LEV		A. Leveque, CA	147	MSCO		S. Miller, AZ	1
LVY		D. Levy, AZ	23	MZS	03	A. Mizser, Hungary	278
LIW		W. Liller, Chile	9	MCE		E. Mochizuki, Japan	21
LMK		M. Linnolt, HI	388	MRV		R. Modic, OH	18
LCO		C. Littlefield, IN	19	MHH		J. Moehlmann, PA	628
LSZ		S. Liu, China	16	MQE		K. Mogul, GA	2412
LLZ	03	L. Liziczai, Hungary	236	MOD		D. Mohrbacher, OH	67
LTE		T. Lloyd Evans, England	1585	MPV	03	P. Molnar, Hungary	113
LOB	06	J. Lobo-Rodriguez, Spain	635	MOZ		Z. Molnar, Romania	9
LRY		R. Lorenz, AZ	9	MLF	10	L. Monard, South Africa	4638
LRD		D. Loring, UT	844	MMOI		M. Montero Reyes Ortiz, Bolivia	17
LDS	20	D. Loughney, England	2	MXO		C. Montes, Philippines	4
LMJ	17	M. Luostarinen, Finland	375	MJOH	20	J. Moore, England	892
MAMB		A. Maasho, TN	2	MEV	01	E. Morelle, France	66655
MDW	27	W. MacDonald, Canada	4901	MOI	01	E. Morillon, France	628
MDD		P. Madden, LA	24	MOW		W. Morrison, Canada	4980
MMT	17	M. Maenpaa, Finland	9	MPS	27	P. Mozel, Canada	88
MLI		L. Maisler, NY	51	MMH		M. Muciek, Poland	23
MDAV		D. Majors, CA	37	MBQ		B. Mullin, MN	11
MVO	17	V. Makela, Finland	85	MJV		J. Murray, OH	11
MUQ		D. Manousos, Greece	1	MUY	05	E. Muyllaert, Belgium	3108
MKE		B. Manske, WI	2	MGW		G. Myers, CA	1031
MGK		G. Maravelias, Greece	216	NDQ	01	D. Naillon, France	286
MRMA		R. Maravillas, NJ	2	NDA		D. Nance, AL	1
MBOA		B. Marinov, Bulgaria	11	NLX	14	P. Nelson, Australia	8318
MFB	01	F. Mariuzza, Italy	101	NAL	03	A. Nemes, Hungary	145
MKW		A. Markiewicz, Poland	749	NBB		B. Neuman, VT	3
MMN	18	M. Martignoni, Italy	86	NAR		A. Neumann, NC	1
MYC		C. Martin, NE	7	NJO	02	J. Neumann, Germany	1851
MDBA		D. Martin, AZ	2	NMI		M. Nicholas, AZ	70
MMG		M. Martinengo, Italy	12	NMR	20	M. Nicholson, England	307
MRX	02	H. Marx, Germany	819	NHS	11	H. Nielsen, Denmark	33
MQI		M. Matesic, Croatia	29	NFD	04	F. Nieuwenhout, Netherlands	214
MTH		H. Matsuyama, Australia	11432	NMT		M. Nissinen, Finland	120
MFE	13	C. Mattos, Brazil	5	NCH		C. Norris, TX	31
MPR	30	P. Maurer, Germany	376	NAO		A. Novichonok, Russia	11
MGE		G. Mavrofridis, Greece	2530	NKL		K. Nuber, Germany	191
MAZ		M. Mazurek, AZ	9	NAN		A. Nygaard, England	65
MBE		B. McCandless, MD	676	OCX		L. O'Connor, MA	39
MUE		R. McDaniel, TX	440	OCN	27	S. O'Connor, Bermuda	115
MDP	27	P. McDonald, Canada	1523	ONJ		J. O'Neill, Ireland	90
MGH	20	H. McGee, England	1099	OSN		S. Oatney, KS	101
MKSA		K. Meagher, MD	4	OWJ		W. Obuchowicz, Poland	14
MEP		D. Medicis, NY	272	OES		D. Oesper, WI	3
MED	20	K. Medway, England	1683	OSL		S. Ogalde, Chile	15
MHI		H. Menali, MA	71	OYE		Y. Ogmen, North Cyprus	14178
MJLE		J. Menke, MD	23	OAR	17	A. Oksanen, Finland	9702
MQB		N. Mennekens, Belgium	3	OPF		P. Olver, Australia	34
MZK		K. Menzies, MA	14148	OAD		A. Ormsby, MI	87
MBO	28	I. Merhebi, Lebanon	3	OJJ		J. Ott, CO	1722
MXH		J. Meriaux, CA	1	OCR	05	C. Otten, Belgium	290
MEZ	03	C. Mezosi, Hungary	16	OMIA		M. Overacker, VA	4
MTK		T. Michalik, VA	39	ORAA		R. Owen, NC	11
MOK	08	O. Midtskogen, Norway	118	OEH		E. Ozturk, Turkey	58
MVH		V. Mihai, Romania	135	PSD		S. Padovan, Spain	964
MDWA		D. Miles, MD	1	PLN		L. Pagel, Germany	154
MXL	20	R. Miles, England	84	PLP		L. Palazzi, Italy	1202

Table 3. AAVSO Observers, 2008–2009, cont.*

Code	Org.	Name	No. Obs.	Code	Org.	Name	No. Obs.
PPS	03	S. Papp, Hungary	3062	RFC		F. Rodriguez Bergali, Spain	390
PREA		R. Paret, France	1	RMU	06	M. Rodriguez Marco, Spain	852
PTQ		T. Parson, MN	1619	ROE		J. Roe, MO	12682
PCG		J. Pascual Gutierrez, Spain	4	RES		E. Romas, Russia	3
PJJ	15	J. Pastor, Spain	2	ROG		G. Ross, MI	210
PKV		K. Paxson, TX	538	RGN		G. Rossi, Italy	40
PTX		T. Peairs, VT	14	RAFA		A. Roussell, Canada	14
PKL		K. Pearson, VA	7	RCJA		C. Roussell, Canada	28
PBT		R. Pearson, VA	12	RR		R. Royer, CA	1
PEI	11	E. Pedersen, Denmark	86	RGY		G. Rubright, PA	7
PEG	01	C. Peguet, France	623	RJV	30	J. Ruiz Fernandez, Spain	194
PWD		W. Pellerin, TX	21	RPH		H. Rumball-Petre, CA	1
PGWA		G. Peterson, Canada	41	RTH		T. Rutherford, TN	119
PHB		H. Peterson, RI	1	RUJ		J. Ruthroff, IN	93
PVA	27	V. Petriew, Canada	1	RZM		M. Rzepka, Poland	2097
PXR	20	R. Pickard, England	3508	SJD		J. Sabia, PA	72
PKI		O. Piechowski, KY	29	SRIC		R. Sabo, MT	12404
PROC		R. Pieri, France	11	SJQ		A. Sajtz, Romania	370
PUWA		U. Pilz, Germany	1	SSU		S. Sakuma, Japan	1143
PLQ	01	L. Pinatelle, France	372	SJAV		J. Salas, Spain	3
PGU	18	G. Pinazzi, Italy	18	SVI		M. Sallman, MN	611
PPL		P. Plante, OH	218	SQL	26	R. Salvo, Uruguay	1
PAW		A. Plummer, Australia	4518	SAH		G. Samolyk, WI	52496
AST	12	R. Podesta, Argentina	33	SXY		A. Sankowski, Poland	3
PRX		R. Poklar, AZ	5618	SGX	03	G. Santa, Hungary	245
PMO	10	M. Poll, South Africa	33	STC		G. Santacana, PR	17
PMV		M. Popescu, Romania	72	ASN		A. Santerne, France	1
PWR		R. Powaski, OH	11	SSIM		S. Santini, Italy	1
PSEA		S. Powers, CA	2	SGE	27	G. Sarty, Canada	10
POX		M. Poxon, England	469	STMA		T. Sauer, Germany	693
PYG		G. Poyner, England	12044	SVA		A. Saw, Australia	174
PAH		A. Price, MA	9	SDAV		D. Scanlan, England	156
POB		R. Price, England	83	SDY	02	D. Scharnhorst, Germany	75
PMB		M. Prokosch, TX	18	SFS		S. Schiff, VA	50
PUJ	06	F. Pujol-Clapes, Spain	683	SPK	01	P. Schmeer, Germany	9
PKU		K. Pukero, Finland	90	SUF		C. Schneider, CA	34
PHG		H. Purucker, Germany	341	SKEA		K. Schneyer, RI	1
QJK	03	J. Qvam, Norway	36	SQE		R. Schoenstene, IL	12
RJB		J. Rachlin, MA	25	SFRA		F. Schorr, GA	504
RKE	02	K. Raetz, Germany	502	SGLE		G. Schrader, Australia	86
RWA		W. Rauscher, PA	25	SYU	02	M. Schubert, Germany	524
RRD	14	R. Rea, New Zealand	77	SAND	02	A. Schumann, Germany	63
RFA		F. Reichenbacher, AZ	54	SRIH		R. Schwartz, WA	4793
RZS	03	Z. Reiczigel, Hungary	4	SJEA	01	J. Sciolla, France	22
REP	24	P. Reinhard, Austria	583	SRYA	27	R. Scott, Canada	46
RFP	13	P. Reis-Fernandes, Brazil	13	SANI		A. Semien, LA	16
RMQ		M. Reszelski, Poland	28	SSHA		S. Shaffer, WY	108
RNA	03	N. Rezsabek, Hungary	4	SHS		S. Sharpe, Canada	3234
RJG		J. Ribeiro, Portugal	2234	SDP		D. Sharples, NY	11
RBJ		J. Richards, Wales	6	SSA		A. Sharpless, WA	18
RNO		N. Richardson, GA	4	SFY	20	J. Shears, England	4753
RIJ		S. Riley, CT	62	SHW		W. Sherman, TX	159
RRJ		R. Rios, CA	2	SLH		L. Shotter, PA	1177
OJR	30	J. Ripero Osorio, Spain	1894	SUY		A. Shoup, OH	1000
RIV		M. Rivera, Italy	164	SSER		S. Shurpakov, Belarus	18
RLJA		L. Robert, France	9	SJAK		J. Sibielak, Ireland	1
RCW		C. Robertson, KS	399	SRWA		R. Sikes, MA	2
RZD	06	D. Rodriguez, Spain	31	SPAO	18	P. Siliprandi, Italy	401
RHE	26	H. Rodriguez, Uruguay	244	SGEO		G. Silvis, MA	58

Table 3. AAVSO Observers, 2008–2009, cont.*

Code	Org.	Name	No. Obs.	Code	Org.	Name	No. Obs.
SNE		N. Simmons, WI	398	SWV		D. Swann, TX	410
SXN		M. Simonsen, MI	3538	SSW		S. Swierczynski, Poland	2428
SANG		A. Sing, Philippines	130	SAO	03	A. Szauer, Hungary	108
SGOR		G. Sjoberg, MA	6117	SXB		M. Szczerba, Poland	1
SDN		D. Slauson, IA	8	SILD	03	I. Szeitz, Hungary	1
SALX		A. Smirnov, Russia	14	TUO		U. Tagliaferri, Italy	67
SJX	10	J. Smit, South Africa	176	TSH		S. Taheran, TX	18
SMI		A. Smith, England	9	TDB	27	D. Taylor, Canada	919
SDEW		D. Smith, OK	15	TJOA		J. Taylor, OR	13
SJE		J. Smith, CA	52	TNX	14	N. Taylor, Australia	33
SUI		R. Smith, England	75	TPV		P. Temple, NM	4
STAK		T. Soejima, Japan	9	TJV		J. Temprano, Spain	1179
SKA	16	K. Sokolovsky, Germany	32	TPS	03	I. Tepliczky, Hungary	990
SGYO	03	G. Soponyai, Hungary	365	TFM		F. Teyssier, France	4558
SYP		P. Soron, Canada	50	TTU		T. Tezel, Turkey	27
SOW	17	J. Sorvari, Finland	7	TBRA		B. Tobias, TX	2
SJOS		J. Spampinato, PA	1	TRL		R. Togni, AR	42
SJZ		J. Speil, Poland	2062	TRE		R. Tomlin, IL	122251
SMUS	27	M. Spicer, Canada	2	TST		S. Toothman, OH	2
SC	27	C. Spratt, Canada	54	TVM		V. Torres, Spain	692
SXR	03	M. Sragner, Hungary	10	TAV	03	A. Tozser, Hungary	8
SBL	05	B. Staels, Belgium	33146	TFR		F. Travaglino, Italy	195
SDAR		D. Stanford, CA	240	TWA		W. Travis, MA	23
SVAE		V. Stanimirov, Bulgaria	5	TRF		C. Trefzger, Switzerland	25
STR		R. Stanton, CA	160	TDW		D. Trowbridge, WA	30
SDB		D. Starkey, IN	1392	TJC		J. Truax, MI	6
SALE	09	A. Staroverov, Ukraine	25	TRX		R. Truta, Romania	5
SPET		P. Starr, Australia	5041	TSJ		S. Tsuji, Japan	26
SJAT		J. Starzomski, Poland	3	TXA		A. Tudorica, Romania	2
STAS		T. Stebler, Switzerland	17	TYS		R. Tyson, NY	811
STF		G. Stefanopoulos, Greece	116	UVR		V. Uher, PA	3
STI		P. Steffey, FL	685	UOS		O. Urquidi, Bolivia	1
SWIL		W. Stein, NM	15474	URS		R. Uyematsu, FL	5
SVR		R. Stencel, CO	1	VFR	01	F. Vaclic, Czech Republic	7
SET		C. Stephan, FL	1022	VLN	01	L. Vadrot, France	104
SBAR		B. Stepinski, Poland	11	BVE	04	E. Van Ballegoij, Netherlands	2324
SJNO	03	J. Stickel, Hungary	251	VBR		H. Van Bemmel, Canada	92
SRB		R. Stine, CA	1105	VDE	04	E. Van Dijk, Netherlands	8
SOX		C. Stockdale, Australia	4433	VNL	05	F. Van Loo, Belgium	1161
STQ		N. Stoikidis, Greece	208	VSH	05	H. Van Sebroeckx, Belgium	28
SDI	20	D. Storey, England	253	VUG	04	G. Van Uden, Netherlands	194
SFU	14	M. Streamer, Australia	75	VWS	05	J. Van Wassenhove, Belgium	281
SNJ		N. Stritof, Slovenia	15	VWA		A. Van Werven, FL	2
SHZ	02	H. Struever, Germany	9	VBH	05	H. Vandenbruaene, Belgium	9
SRX	14	R. Stubbings, Australia	9120	VAEA		A. Vandusen, IN	2
SUK		M. Stuka, CA	6	VSD	05	D. Vansteelant, Belgium	24
SUS	02	D. Suessmann, Germany	695	VKN		K. Vardijan, Croatia	2
SJAR	17	J. Suomela, Finland	505	VMG		M. Vargas, Portugal	2

** Totals do not include historical data (data preceding fiscal 2008–2009) submitted during 2008–2009.*

Table 3. AAVSO Observers, 2008–2009, cont.

These codes, which appear in the Table (AAVSO Observers 2007–2008), indicate observers are also affiliated with the groups below:

01 Association Française des Observateurs d'Étoiles Variables (AFOEV)
02 Bundesdeutsche Arbeitsgemeinschaft für Veränderliche Sterne e.V. (BAV) (Germany)
03 Magyar Csillagàszati Egyesület, Valtózocsillag Szakcsoport (Hungary)
04 Koninklijke Nederlandse Vereniging Voor Weer-en Sterrenkunde, Werkgroep Veranderlijke Sterren (Netherlands)
05 Vereniging Voor Sterrenkunde, Werkgroep Veranderlijke Sterren (Belgium)
06 Madrid Astronomical Association M1 (Spain)
08 Norwegian Astronomical Society, Variable Star Section
09 Ukraine Astronomical Group, Variable Star Section
10 Astronomical Society of Southern Africa, Variable Star Section
11 Astronomisk Selskab (Scandinavia)
12 Liga Ibero-Americana de Astronomia (South America)
13 Brazilian Observational Network REA
14 Royal Astronomical Society of New Zealand, Variable Star Section
15 Agrupacion Astronomica de Sabadell (Spain)
16 Association of Variable Star Observers "Pleione" (Russia)
17 URSA Astronomical Association, Variable Star Section (Finland)
18 Unione Astrofili Italiani (Italy)
20 British Astronomical Association, Variable Star Section
21 Israeli Astronomical Association, Variable Star Section
23 Grupo Astronomico Silos (Spain)
24 Astronomischer Jugendclub (Austria)
26 Red de Observadores (Montevideo, Uruguay)
27 Royal Astronomical Society of Canada
29 Asociacion Amigos de la Astronomia (Argentina)

Table 4. Observation statistics for fiscal year 2008–2009.*

Observations (increments of 1000)	*No. Observations per increment*	*% of All Observations*	*No. Observers per increment*
1-999	82256	7	628
1000-1999	53796	5	37
2000-2999	45369	4	19
3000-3999	38282	3	11
4000-4999	63492	6	14
5000-5999	28098	3	5
6000-6999	19448	2	3
7000-7999	22165	2	3
8000-8999	8318	1	1
9000-9999	28173	3	3
10000+	707958	65	21

** Totals do not include historical data (data preceding fiscal 2008–2009) submitted during 2008–2009.*

Section Reports

Cataclysmic Variable (CV)

Section Leaders: ***Mike Simonsen,*** *2615 S. Summers Road, Imlay City, MI 48444*
Gary Poyner, *67 Ellerton Road, Kingstanding, Birmingham, B44 0QE, England*

As part of the committee restructuring process initiated by the AAVSO Council in 2009, a section devoted to the observation and study of cataclysmic variables was created in April, 2009. The section leaders responsible for administering the section are Mike Simonsen and Gary Poyner. Dr. Boris Gaensicke and Dr. Paula Szkody graciously agreed to serve as scientific advisers for the section.

A section website was built and launched April 1, 2009, and has proved to be a success in the first eight months. News items and recently published papers are featured on the home page, as well as a continuously updated listing of CV outbursts reported in the previous seventy-two hours called Activity at a Glance. There is also a running tally of CVs brighter than 16th magnitude detected by the Catalina Real Time Survey. *AAVSO Alert Notices* and *Special Notices* pertaining to CV activity are published here, as well as relevant *IAU Circular* information and notices from the *Astronomer's Telegrams*.

Several campaigns or observing projects are featured, with links to the lists and web pages devoted to them. These include the Hamburg Survey CV list published by Gaensicke, the BAAVSS Long Term Polar Monitoring Programme, and the CV Section's latest observing project, The Z Campaign, designed to study Z Cam type dwarf novae.

Articles and interviews with CV observers and researchers are published online via the CV Blog and Articles pages, and observers are kept abreast of discussion and activity through our email lists, CVnet-discussion, CVnet-outburst, and CVnet-circular.

With hundreds of people subscribed to the mail lists and over 12,000 hits to the website from 84 countries worldwide since May 1, 2009, the CV Section is off to a fine start. We hope to build on this early success and improve the services, programs, and website as time allows.

Data Mining

Section Leader: ***Michael Koppelman,*** *1523 Valders Avenue N, Golden Valley, MN 55427*

The AAVSO Data Mining section was created by the AAVSO Council at the Fall 2008 Meeting. I was appointed section leader by the Council at the Spring 2009 meeting. I gave a brief report at the Spring 2009 Meeting based on my discussion with the Director about the general direction of the section. There are three main goals of the section:

1. Provide education and training in data mining.
2. Support and direct original data mining research.
3. Aid existing programs with data mining tasks.

In the months between the Spring Meeting 2009 and the Annual Meeting 2009 a few basic apparati were put in place to enable and encourage section participation by the membership:

- A web site was created at http://datasection.aavso.org/. It is functionally a sub-site of the main AAVSO site but for now is hosted separately at Google. The web site contains basic, static information about the section.

- A group discussion area at http://groups.google.com/group/aavsodatasection. The group currently has about thirty members, including some AAVSO staff and myself. Discussion activity has been of fairly low quantity but pretty high quality. There is plenty to sink one's teeth into. There are currently fifty-four messages in twenty different discussion topics.

- We signed up one scientific advisor (Dr. Doug Welch). Patrick Wils declined a scientific advisor role but said he was happy to be a member and mentor of sorts, which he has been.

There is tons of great data mining science going on and we hope to support and encourage AAVSO members to participate in it. My job as section leader is to organize and report on the activities of the section. Scientists like Doug and participants like Patrick are here to provide ideas and advice. What this group accomplishes is ultimately up to the members. We welcome and encourage all members to get involved.

Eclipsing Binary

Section Leaders: ***Gerard Samolyk,*** *P.O. Box 20677, Greenfield, WI 53220*
Gary Billings*, P.O. Box 263, Rockyford, Alberta T0J 2R0, Canada*

Gerry Samolyk and Gary Billings are leading the newly organized Eclipsing Binary Section. Gary has started work on a new website for this section. Dr. Dirk Terrell, who has been our advisor for a number of years, has stepped down. We thank him for his many contributions, especially for his 2001 workshop that introduced us to modelling binary systems using the Wilson-Devinney code. Dr. Ed Guinan has agreed to serve as advisor for the Section.

Starting in 2009, the ephemeris was expanded to include a total of 200 stars. AAVSO observers have closely monitored all of these stars for thirty years or more. The ephemeris can be found at:

http://www.aavso.org/observing/programs/eclipser/ebephem.shtml

Two papers containing a total of 484 times of minima have been submitted to *The Journal of the AAVSO* for publication. In addition, a paper about the recent eclipse of the long period star EE Cep was published in *JAAVSO*. Times of minima published by the AAVSO continue to be added to the Lichtenknecker Database maintained by the Bundesdeutsche Arbeitsgemeinschaft für Veränderliche Sterne e. V. (BAV). An English language interface to this database can be found at:

http://www.bav-astro.de/LkDB/index.php?lang=en.

The project to upload the visual observations made from 1975 thru 2000 to the AAVSO International Database is close to completion. With the exception of a few problems, all of the observations from the old Apple format have been translated and uploaded. For visual observations made using steps rather than magnitudes, the step value is located in the comment field of the database record.

The rare eclipse of ε Aur started in August of 2009 and will continue for the following two years. For more information, refer to the AAVSO Citizen Sky Project at: http://www.citizensky.org/

Prepared by Gary Billings

Education and Outreach

Section Leader: ***Pamela Gay****, Southern Illinois University Edwardsville, Physics Dept., Box 1654, Edwardsville, IL 62026*

In the face of the International Year of Astronomy (IYA), the members of the AAVSO Education Committee have been more than a little busy trying to spread astronomy and variable star astronomy around the globe through their local and national IYA organizations.

Several committee members have worked on their own on projects that need recognizing. Specifically, I'd like to recognize Mario Motta, M.D. for his work with the American Medical Association to introduce and pass a policy for AMA to support light pollution laws nationwide. Thanks to Mario's efforts, activists trying to get state or local laws passed can now reference the AMA policy to support keeping our skies dark.

We have also seen two members of our committee, longtime member John Percy and new member Arif Solmaz, use the IYA as a platform to speak about variable stars, with John using an ASP SEED grant to bring Galileoscopes into a modified astronomy and optics curricula which, when possible, will facilitate students' observing variable stars. Arif has also directed his observatory in Canakkale, Turkey, and its numerous public nights toward observing variable stars.

Beyond these individual efforts, the group efforts of four projects lead by Mike Simonsen are particularly worth noting. These activities are the AAVSO Speakers Bureau, Writers Bureau, and Presentation Library. Since these three projects were launched at the March 2008 meeting, twenty-two people—professionals and amateurs—have joined the AAVSO Speakers Bureau, and this year this group reports giving over 100 presentations around the world, with more than half of these talks and workshops focusing specifically on variable stars. More than half of the talks reported were given by five people: Jaime Garcia, John Percy, Arif Solmaz, Mike Simonsen, and myself; and I'd like to thank Jaime, John, Arif, and Mike for giving so much of their time and energy to speak on variable stars and astronomy.

The Speakers Bureau is always looking to grow. If you are interested in joining, please see Mike or me at any time.

To help you give talks with less effort, we have worked to create a library of presentations. Through the AAVSO website, you can access fourteen presentations from eight different people, including a presentation on visual observing available in both English and Spanish.

If you have a presentation you think may help others communicate aspects of variable star astronomy and observing to others, please talk to Mike or me about how to get your presentation online.

Our final online resource is the AAVSO Writers Bureau. This site, curated by Mike Simonsen with help from Arif and Ken Mogul, aggregates astronomy stories from around the Blogisphere and makes them available to astronomy clubs and societies to use in their newsletters. This is a password protected site; to receive access, ask Mike Simonsen. Used regularly by over fifty editors, the stories by the eighteen contributing writers have appeared hundreds of times in newsletters throughout the English-speaking world. If you have empty space in your club's newsletter, this AAVSO service may be your answer.

The final area we have worked hard in is the AAVSO Mentoring program. In the past year, thirty-one mentors have helped fifty-three mentees better learn how to contribute to variable star astronomy. People assisted by this program have ranged from beginning astronomy newbies through to professionals needing help learning how to get the most out of their classroom equipment. If you are interested in becoming a mentor or are new and would like a mentor, ask Mike.

One recurrent theme through this report is "ask Mike." This year the mentoring and outreach efforts of the AAVSO were largely driven by the efforts of Mike Simonsen. Without him, many of us would have had a poorer year. Thank you Mike, and many thanks to all of you who have given in to his arm-twisting. All your efforts have made an impact in communicating astronomy to other amateurs and the general public.

Thank you.

Long Period Variable (LPV)

Section Leaders: ***Katherine Hutton***, *969 E. Mountain Street, Pasadena, CA 91104*
Mike Simonsen, *2615 S. SUmmers Road, Imlay City, MI 48444*

In 2008, the AAVSO restructured its existing committees into a number of Sections, according to the type of variable star of interest. This report comes from the thus newly-formed Long Period Variable Section (LPV). The LPV Section committee consists of Mike Simonsen, Michael Koppelman, and myself. Our scientific advisors are Dr. John Percy, Dr. Laszlo Kiss, Dr. Lee Ann Willson, and Dr. Matthew Templeton. We include in our purview the following types of variable stars: Mira (M, M:), SR (including SRA, SRB, SRC, SR:, etc.), all the L-type stars (L, LC, etc.), and the RV Tau stars.

Although other variable star types have received much attention, the long period variables (LPVs) have been the "bread and butter" of the AAVSO observing program since its inception in 1911. The AAVSO database for LPVs consists primarily of visual observations, augmented by PEP, and recently, CCD observations, with a cadence from one observation per week to one per month, per observer. A few stars have over 100 years of continuous observations. Over time, the number of known LPVs has increased vastly, the number and geographic location of observers has changed, charts and the availability of information have greatly improved, and instrumentation has undergone a revolution (or two).

Because large automated surveys—ASAS being one example—now or will soon make more and more accurate observations of long period variable stars than the large and dedicated team of AAVSO observers could ever make, it has been necessary to reconsider the list of LPVs recommended for observation. Stars were evaluated based on: 1) their usefulness as training stars for new observers, 2) the number of observations already in the database, 3) the number of scientific citations given by the SIMBAD listing, and 4) special requests from professional or amateur observers.

We now have four lists of recommended stars:

1) 10 training stars,
2) 100 AAVSO "Legacy" stars, with the most AAVSO data over the longest time period already in place,
3) 390 AAVSO "Program" stars (including the above), each with more than 5,000 observations already in place,
4) "Binocular" stars, which are too bright to be included in ASAS-type surveys.

These lists, plus other pertinent information, can be found on the Section web page: http://sites.google.com/site/aavsolpvsection/. There is a link under "Section Sites" on the left-hand frame of the main AAVSO home page.

Observers are encouraged to continue observing any stars already part of their program, or which particularly interest them. Observers looking for suggestions are encouraged to select from the Legacy, Program, or Binocular lists, depending on their sky conditions and equipment. CCD observations with astrometric filters are especially encouraged, but visual observations are still welcomed as well. These program lists are dynamic, subject to special requests from astronomers. The Sequence Team has been doing a yeoman's task of updating the sequences for the LPV's. So—enjoy the night sky and don't forget to submit the results of your observations through WebObs!

Nova Search

Section Leader: ***Reverend Kenneth C. Beckmann,*** *330 North Washington, Kahoka, MO 63445*

From September 1, 2008, through August 31, 2009, the following novae were discovered:

V1309 Sco (Nova Scorpii 2008) was independently discovered on September 2, 2008, by K. Nishimaya and F. Kabashima of Japan; by Y. Sakurai of Japan; and by Guoyou Sun andXing Gao of China.

V1721 Aql (Nova Aquilae 2008) was discovered on September 22, 2008, by K. Itagaki of Japan.

QY Mus (Nova Muscae 2008) was discovered on September 28, 2008, by William Liller of Chile.

V679 Car (Nova Carinae 2008) was discovered on November 26, 2008, in the Pi of the Sky Automated Survey database and reported by K. Marliak of Poland.

V5580 Sgr (Nova Sagittarii 2008 No. 2) was discovered on November 29, 2008, by William Liller of Chile.

Nova LMC 2009 was discovered on February 5, 2009, by William Liller of Chile.

V5582 Sgr (Nova Sagittarii 2009 No. 2) was discovered on February 23, 2009 (but not announced until May 29), by G. Sun and G. Gao of China.

V5581 Sgr (Nova Sagittarii 2009 No. 1) was discovered on April 21, 2009, by K. Nishiyama and F. Kabashima of Japan.

Nova LMC 2009 No. 2 was discovered on May 4, 2009, by William Liller of Chile.

V1213 Cen (Nova Centauri 2009) was discovered on May 8, 2009, in the ASAS database by G. Pojmański, D. Szczygiel, and B. Pilecki of Poland.

V5583 Sgr (Nova Sagittarii 2009 No. 3) was independently discovered on August 6, 2009, by K.Nishiyama and F. Kabashima of Japan; and in the ASAS database by G. Pojmański, D. Szczygiel,and B. Pilecki of Poland.

V2672 Oph (Nova Ophiuchi 2009) was discovered on August 16, 2009, by K. Itagaki of Japan.

We congratulate all those who discovered novae during the past year.

Interest has been shown in updating the Atlas of Historical Novae which appears on the Nova Search web pages of the AAVSO website. Originally, the plan was to update this atlas every five years. It is our hope that the atlas will again be current in 2011.

During the past year, from September 1, 2008, until August 31, 2009, the following observers participated in the program:

Kenneth Beckmann	USA	267	Gary Nowak	USA	1,287
Manfred Durkefälden	Germany	24	Richard Wobus	USA	18

We thank these observers for their participation in the program and encourage others to participate and send their observations.

Photoelectric Photometry

Section Leader: ***James H. Fox,*** *P.O. Box 135, Mayhill, NM 88339*

The AAVSO Photoelectric Photometry (PEP) program had a fantastic year in 2009, with new observers joining the program and more contributions by our long-term PEP observers. For the fiscal year, 940 PEP observations were submitted through the PEPObs utility of WebObs, and an additional 1,178 observations were submitted as reduced magnitudes, for a total of 2,118 observations. This is an increase of over 60 percent over FY 2008's total of 1,317 observations! The total also includes 333 PEP-*IR* observations made in the *J* and *H* bands of the infrared.

During the year, we added several stars to the PEP program, including the Cepheids δ Cep and Polaris, the high-mass X-ray binary V884 Sco, β Cep, and the pulsating red giants V2105 Oph and χ Peg. Along with these have come campaigns on several of these stars as well as the luminous blue variable P Cyg, and ε Aurigae, which began its once-in-27 years eclipse during August of this year.

An exciting new development has been the addition of archival data sets from a few observers and observing groups during the past year. These included: archival *UBV* photometry of ε Aurigae by Jeff Hopkins; archival *UBV* photometry of southern Miras and other stars by the Auckland Photometric Observers Group (headed by Stan Walker); archival *UBV* photometry of several RV Tauri stars from the 1950s by Frank Kameny; and as of today (November 25), over 5,500 photoelectric *UBVRI* observations of southern Miras by Leopoldo Celis, digitized by Brian Skiff of the Lowell Observatory. Not including the Celis data which were added today, more than 13,000 photoelectric observations were added to the AAVSO International Database during 2009!

Photoelectric data are highly prized by researchers. These data are fully calibrated and transformed, and are some of the highest-precision and highest-accuracy data in the AAVSO archives. Furthermore, photoelectric observers are capable of observing the brightest stars in the sky with much greater ease than imaging systems, providing important data on these sorely neglected variables. Photoelectric photometry is painstaking and precise work. A single magnitude may take nearly half an hour to obtain, with multiple measures taken of the variable, comparison star, and sky to create a single data point. The resulting light curves, however, are remarkable and are well worth the effort put into making them.

I'd like to take this opportunity to thank all of our photoelectric observers for their hard work over the past year—thank you for making this a successful year for the AAVSO Photoelectric Photometry program!

PEP Observers for FY 2008–2009 (provisional totals):

Name	*Observations*	*Name*	*Observations*
Brian McCandless	589	Jim Fox	68
David Williams	390	Adrian Ormsby	64
Nick Stoikidis	199	Robert Crumrine	38
Jeff Hopkins	165	James Wood	37
Gianni Galli	132	Hans Nielsen	33
Charles Calia	113	Thomas Peairs	14
Thomas Rutherford	105	Erik Hoeg	3
Glen Ward	84	Wayne Clark	1
Henri Van Bemmel	83		

Prepared with the assistance of Matthew Templeton.

Short Period Pulsator

Section Leaders: ***David A. Hurdis,*** *76 Harbour Island Road, Narragansett, RI 02882*
Gerard Samolyk*, P.O. Box 20677, Greenfield, WI 53220*

The AAVSO Short Period Pulsator (SPP) Section was begun in 2009. It incorporates the AAVSO's active RR Lyrae Observing Program, but also includes the observation of all pulsating stars found on the Hertzsprung-Russell diagram's "instability strip," namely, the Cepheid variables, their Population-II siblings, the W Virginis stars, RR Lyr stars, and the very short period δ Scuti stars.

The SPP section's co-leaders are Gerry Samolyk and Dave Hurdis, while Shawn Dvorak is its webmaster. Its website can be accessed directly at http://sites.google.com/site/aavsosppsection/, or via a link on the Observing Programs page of the AAVSO website.

Three professional astronomers have kindly agreed to serve as scientific advisors to the SPP Section. They are: Prof. Horace Smith, Michigan State Univ. (RR Lyr); Prof. Doug Welch, McMaster Univ. (Cepheids); and Dr. Matt Templeton, AAVSO Headquarters (Time Series Analysis).

At the AAVSO Annual Meeting on November 7, 2009, Gerry Samolyk presented an overview of the SPP Section. The section's purpose is to provide guidance and useful information to observers interested in short period pulsators. This includes suggestions of stars for which observations are particularly needed. Gerry presented a table of RR Lyrae "legacy" stars needing more observations. Although understanding Cepheid variables is very important for astronomical distance measurement, the AAVSO Cepheid observation program has been inactive recently. The SPP Section hopes to revive interest in observing this important class of stars, and Gerry presented a table of Cepheids that AAVSO observers have observed in the past. The tables of RR Lyr legacy stars and Cepheid stars can be found on the SPP Section website.

Solar

Section Leader: ***Paul Mortfield,*** *34 Portree Crescent, Thornhill, ON L3T 3G2, Canada*

The dedicated group of AAVSO solar observers continue to monitor the sun both visually and in radio wavelengths. These observers should be given credit for keeping vigil in spite of a very prolonged solar minimum. We welcome new observers to our ranks and congratulate the observers mentioned below that have achieved awards for observational milestones.

The chair acknowledges with thanks the work of: Michael Hill, AAVSO SID Analyst; Dan Williams, AAVSO Sunspot Analyst; and headquarters volunteer Arthur Ritchie. The committee also wishes to thank AAVSO staff members Kate Davis, Virginia Renehan, and Richard "Doc" Kinne for their continued assistance and support of our efforts.

Sunspot Observation Group

During the past thirteen months there were 77 observers who contributed a total of 12,309 observations.

Observers with 300 or more observations are:

Brenda Branchett	370	Brian Cudnik	308
Alan Buck	365	David Teske	304
German Morales	362	Franky Dubois	302
Gema Araujo	347	Miyoshi Suzuki	302
Hulya Yesilyaprak	309		

We welcomed the following new sunspot observers:

Alexandru Burda	BXD	Hans-Goran Lindberg	LINH
Pieter-Jan Dekelver	DEKP	Fabio A. Mariuzza	MARF
Sjoerd Dufoer	DFS	John Rousom	RO
Anna Hillier	HILA	Krystyna Wirkus	JASK

Two observers are eligible for awards this year. Sunspot Observer awards are given to observers after having submitted 1,000, 1,500, or 2,000 monthly reports. The observers are:

Patrick Abbott	AAP	Howard Barnes	BARH

Sudden Ionosphere Disturbance Group

For the last twelve months SID Activity has been very slow to nonexistent. We continue to experience a very prolonged time with minimal flare activity on the sun resulting in only four correlated SID events over the course of the year. Our observer ranks have remained consistent and we still have a good number of observers remaining vigilant in their watch for the next solar flare events. We even added two new observers. There were a total of seventeen observers submitting reports and a total of 157 reports were sent in. Thanks to all observers for their efforts in monitoring, data analysis and report generation.

Six observers are eligible for awards this year. SID Observer awards are given to observers after having submitted 40 reports to the group. The observers are:

Andy Clerkin	A29	Roberto Battaiola	A96
Peter King	A80	Jon Wallace	A97
Michael Hill	A87	Lionel Loudet	A118

Supernova Search

Section Leader: AAVSO Headquarters

The AAVSO Supernova Search Section is under development. Section Leaders, programs, and procedures will be announced soon!

Treasurer's Report
October 1, 2008–September 30, 2009

Gary W. Billings, *Treasurer, AAVSO, 49 Bay State Road, Cambridge, MA 02138*

The financial figures provided herein are, as in previous years, unreconciled and unaudited, as available a few weeks after fiscal year-end, and in time for the Annual Meeting. In accordance with legal requirements, AAVSO has its "books" audited yearly by an external accountant, but the auditor's report is not available at the time of this writing.

The following comments are offered to aid in interpreting this report. Except for comments specifically mentioning 2009, these apply to previous years' reports as well.

The Income section does not just list monies that are new to the AAVSO. For example, it includes money transferred from our "endowment" accounts. In light of recent declines (negative returns) in the stock markets, in 2009 some of this "income" is actually drawn from the principal of those accounts, not just from interest and appreciation. Such withdrawals, and the level of risk associated with those withdrawals, are subject to a great deal of careful deliberation by Council.

Of note in Income is the high level of donations, bequests, and grants in 2009. Arne and Linda Henden have been most generous, and we continue to receive money from the Dorrit Hoffleit bequest. In 2009 we obtained a significant grant from the NSF to execute the Citizen Sky outreach project. This grant will continue for the next several years. We also received numerous grants from individuals to establish AAVSOnet—a network of robotic telescopes.

The Expenses section likewise needs more explanation. Most significantly, it does not include spending towards the purchase or upgrading of capital assets. In this case, it omits almost $30,000 of renovations that increase the value of our headquarters building, and about $22,000 of equipment for AAVSOnet.

Some further items that must be considered when comparing Income to Expenses are changes in cash on hand at year end, and changes in the amount of receivables and prepaid expenses (mostly insurance). In FY 2009, because of when grant monies arrived, we drew more from our endowments that we actually needed to. This sum was "covered" by grant money received just before fiscal year end, leaving us with a lot of cash on hand.

Thus, while one might look at Income, which exceeds Expenses by a large sum, and conclude we generate a large surplus every year, that is not the case!

To put some numbers to these explanations, I have added a small section to this report, showing how these adjustments apply. Note that these are preliminary numbers, unaudited, and don't actually balance (yet). With that caution, I include them to indicate the general, but not exact, financial picture.

2009 Income

Dues income	$57,863
Sales	11,567
Meetings	6,208
Grants	145,210
Bequests and Donations	168,587
Transfers from endowments	890,976
Bank interest	309
Total Income	$1,281,330

2009 Expenses

Staff salary costs	$723,914
Contract/temp salaries	14,895
Payroll tax, benefits	183,161
Building maintenance	1,666
Utilities, cleaning, insurance	20,527
General office expenses	10,734
Postage	16,961
Legal and accounting	16,693
Publications	19,406
Technical operations (including AAVSOnet)	39,132
Internet	7,627
Meetings (including CitizenSky)	51,397
Travel	20,587
Miscellaneous	11,927
Total Expenses	$1,138,627

2009 Disposition of Income (Preliminary)

Total Income	$1,281,330
Total Expenses	(1,138,627)
Additions to buildings	(29,995)
Purchases of equipment	(21,979)
Adjustment for equipment donation in kind	1,343
Change in prepaid expenses	(1,325)
Change in payables	(7,783)
Change in year end cash on hand	(77,402)
Discrepancy (to be resolved)	$5,562

3. Officers, Staff, and Volunteers

AAVSO Officers, Council Members, and Section Leaders for Fiscal Year 2009–2010

You may contact these persons through AAVSO Headquarters.

Officers

Director	Dr. Arne Henden	(term of office: 2005–2010)
President	Jaime Ruben Garcia	(2009–2010)
Past President	Dr. Paula Szkody	(2009–2010)
Vice President	Mike Simonsen	(2009–2010)
Secretary	Gary Walker	(2009–2010)
Treasurer	Gary Billings	(2009–2010)

Council Members

Dr. Pamela L. Gay	(2009–2011)
Dr. Edward F. Guinan	(2008–2010)
Dr. Katherine Hutton	(2008–2010)
Michael Koppelman	(2008–2010)
Dr. Arlo U. Landolt	(2008–2010)
Dr. Jennifer Sokoloski	(2009–2011)
Dr. David Turner	(2009–2011)
Christopher Watson	(2009–2011)

Section Leaders

Cataclysmic Variable	Mike Simonsen, Gary Poyner
Data Mining	Michael Koppelman
Eclipsing Binary	Gerard Samolyk, Gary Billings
Education and Outreach	Dr. Pamela L. Gay
Long Period Variable	Dr. Katherine Hutton, Mike Simonsen
Nova Search	Rev. Kenneth C. Beckmann
Photoelectric Photometry	James H. Fox
Solar	
Section Leader	Paul Mortfield
Sunspot Group Leader	Dan Williams
Solar Flare/SID Observing Group	William Michael Hill
Short Period Pulsator	David Hurdis, Gerard Samolyk
Supernova Search	AAVSO Headquarters
Journal of the AAVSO Editor	Dr. John R. Percy

AAVSO Headquarters Staff

Sara J. Beck	Technical Assistant, Special Projects
Gloria Ortiz Cruz	Data Entry Technician, Part Time
Katherine Davis	Astronomical Technical Assistant, Website
Dr. Arne Henden	Director
Linda Henden	Administrative Assistant, Part Time
Richard Kinne	Astronomical Technical Assistant, Technology
Kerriann Malatesta	Astronomical Technical Assistant
Gamze Menali	Astronomical Technical Assistant, *Newsletter* Editor
Aaron Price	Astronomical Technical Assistant, Technology
Virginia Renehan	Administrative Assistant, Publications
Arthur Ritchie	Headquarters Volunteer
Dr. Michael Saladyga	Technical Assistant, *JAAVSO* Production Editor, Archives, Library
Mike Simonsen	Development Director
Dr. Matthew Templeton	Staff Astronomer, *JAAVSO* Assistant Editor
Rebecca Turner	Astronomical Technical Assistant, Citizen Sky Project Manager, Meeting Coordinator, Part Time
Elizabeth O. Waagen	Senior Technical Assistant, *JAAVSO* Associate Editor

Volunteer Superstars

Members contribute to the AAVSO in a lot of ways. Supporting the organization financially is not the only thing you can do to help the AAVSO with its mission. We have dozens of volunteers working on important projects, such as The International Variable Star Index (VSX), charts and sequences, and the chart plotter (VSP). Others train observers through the Mentor Program, or do outreach through the Speakers Bureau and Writers Bureau.

A lot of this activity goes on behind the scenes, so I'd like to pull back the curtain and introduce you to some of the volunteers doing all this work. There isn't room to mention them all, but here are some of the exceptional examples, our "volunteer superstars."

Tom Bretl, Plymouth, MN. In addition to being an active contributing observer, Tom has been creating new sequences, revising old sequences, and addressing chart errors reported to the Chart Error Tracking Tool at an amazing rate. He is responsible for at least 75% of all the new chart updates since July 2009.

Tim Crawford, Arch Cape, OR. Tim has been our most active and productive mentor in the mentor program. He has helped a dozen new CCD observers learn how to do everything from point their telescopes to flat fielding to reducing time series data. He has even helped teach the coordinator of the mentor program a thing or two! He has a seemingly unlimited amount of patience and always tells his students, "there is no such thing as a stupid question." Tim is also a member of the charts and sequences team.

Tim and Tom are also members of the AAVSO Speakers Bureau, volunteering to drive hours from home to give talks to scout troops, star parties and astronomy clubs for free.

Ken Mogul, Newnan, GA. Ken came to the AAVSO through the mentor program. Physically unable to go out and observe, we eventually got him a mentor, Bill Dillon and time on Global Rent-A Scope (GRAS). Ken began observing variables with robotic telescopes in his first year of observing. Now he has become a mentor to other new robotic telescope users and was a presenter in the recent Photometrica workshop in Newton, MA, teaching others how to use Photometrica, the new AAVSO photometry software available soon as a member benefit. He has also helped select articles for the AAVSO Writers Bureau blog, an important time-intensive activity.

John Blackwell, Exeter, NH. John is an astronomy educator and director of the Phillips Exeter Academy Grainger Observatory. His experience with CCDs, software, observatory controls, and a vast array of telescopes has proven invaluable to the mentor program. John has taken on students along the east coast and mid-west and is the mentor I send other educators to when they are looking for help.

We recently introduced John to an educator at Radford University, in southwest Virginia, who needed help. Radford U. has a 14.5" RCOS telescope with a SBIG STL-11000 camera, run with TheSky6 and MaximDL. John, who is familiar with the telescope and software, is helping them by teaching the instructor the ins and outs of CCD photometry, particularly taking good flats, a challenge they were struggling with.

Christopher Watson, San Diego, CA. I first came to know Chris through working on the chart team. He developed an online tool for keeping track of all the charts we were preparing and checking which eventually evolved into CHET, the Chart Error Tracking Tool. He also designed the interface for VSP, which many of you use on a regular basis, as well as the interface for the administrative tool for the Variable Star Database (VSD), which I now use on a regular basis to perform updates to the comparison star data that VSP plots charts with.

Chris wrote the program that produces the cataclysmic variable section CV Circular. This program searches the AAVSO database every Monday morning and organizes a document of all the activity reported on every CV in the database that is emailed to a list of subscribers. This and the VSD admin tool are examples of processes that run totally behind the scenes, but are vital to making things work at AAVSO.

Chris' greatest contribution has to be the development of the International Variable Star Index, VSX. Originally designed to be a repository for variable star information, kind of a GCVS on steroids, this tool has become the central engine driving several processes within the AAVSO now. The chart plotter, light curve generator, quick look files, WebObs, Zapper, and data download programs all query VSX before they can perform their tasks. As if all this and a family and career weren't enough, Watson now serves on the AAVSO Council, too!

Sebastian Otero, Buenos Aires, Argentina. Sebastian is well known for his near photometric visual observations and is a prolific observer of southern hemisphere binocular and naked eye variables. He has mentored observers in Mexico, South America, the Philippines, New Zealand, and Australia.

Even more remarkable is the substantial time and talent he has invested into moderating submissions to VSX, and updating the database. He was recently given the AAVSO Director's Award for his contributions in this area.

Patrick Wils, Hever, Belgium. Another Director's Award recipient, Patrick has been instrumental in importing large datasets of variables into VSX and has devoted hundreds of hours of his time to approving submissions and guiding contributors through the submission process.

Without Patrick Wils and Sebastian Otero the International Variable Star Index would be years behind where it is today. Without Chris Watson it wouldn't even exist.

Today the AAVSO has a spacious headquarters, an impressive database, a humungous website, and a multi-million dollar endowment, but the most important asset is, and always has been, the members of the organization. It is my privilege and honor to know and work with so many of you.

Mike Simonsen
Development Director
AAVSO

AAVSO Volunteers

AAVSO members are very generous with their time and talents. Many of the programs and services we offer would not be possible without the participation of member volunteers. They are regularly involved in teaching new observers, writing articles for our publications, vetting submissions to the *Variable Star Index*, and the creation of charts and comparison star sequences.

We take this opportunity to recognize these special people, and to say *thank you* for another year of valuable contributions of time and expertise.

Mentor Program Volunteers

Patrick Abbott
John A. Blackwell
Thom Bretl
Glenn Chaple Jr.
Tim Crawford
Bill Dillon
Jim Fox
Geoff Gaherty
Bill Goff
Keith Graham
Kate Hutton
Michael Koppelman
Tom Krajci
Michael Linnolt
Mike Mattei
Peter Nelson
Sebastian Otero
Chuck Pullen
Steve Robinson
Guido E. Santacana
Mike Simonsen
Ray Tomlin

Variable Star Index *(VSX) Moderators*

Michael Koppelman
Sebastian Otero
Wolfgang Renz
Mike Simonsen
Christopher Watson
Patrick Wils

Charts and Sequences

Thom Bretl
Tim Crawford
Robert Fidrich
Keith Graham
Kate Hutton
Michael Koppelman
Jim Jones
Mati Morel
Peter Nelson
Sebastian Otero
Wolfgang Renz
Mike Simonsen
Dan Taylor

Speakers Bureau

Raymond Benge
Tom Bretl
Glenn Chaple
Tim Crawford
Pamela Gay
Keith Graham
Albert Holm
Kate Hutton
Roger S. Kolman, PhD
Tom Krajci
Doug Lombardi
Alex McConahay
Mario Motta
Gordon Myers
Chuck Pullen
Michael Richmond
Michael Rupen
Mike Simonsen
Arif Solmaz
Chris Stephan
Bob Stine
Paul Temple

Solar Section

William Michael Hill
Paul Mortfield
Arthur Ritchie
Dan Williams

AAVSO Newsletter Contributing Authors

Tim Crawford
Bill Dillon
Gerry Dyck
Albert Jones
Brian McCandless
Susan Oatney
Arto Oksanen
Alan Plummer
Tom Richards
Nikolai Samus
Chris Stephan

4. Word From the Astronomical Community

My master thesis was about to fail because our telescope broke down. I had only three months to finish it and the telescope was not going to be ready by that date. So I REALLY needed data for doing my eclipse mapping.

Thanks to your help and the excellent AAVSO data, I won a scholarship to do my PhD in England. I'm so happy!! Thanks again for all your help with my thesis.

—Penelope Longa
Graduate student
Chile

If you make a report to the Variable Star Index (VSX) all this information is vital, as to have your discovery reported and confirmed it needs to be verifiable by others, and available for peer review. Thus the onus is on the discoverer to present the data that needs to be reviewed.

As this was the first time I had been through the process, and being very much in the AMATEUR Astronomy camp, I was very nervous about the process. Its always important to follow the process and accept the feedback that comes from those with much more experience than one's self.

I received back a very helpful email that rejected my submission (for now) due to the fact that I didn't have enough data to produce a full phase diagram and I had suggested that it may be a Cepheid Variable due to the short period and hadn't considered that it was a bit too blue to be a Cepheid and that it was more likely a RRab. The VSX person made some very helpful suggestions about what I should do next—get some more data and produce a full phase diagram and re-submit.

—Peter Lake
AAVSO member
Australia

The AAVSO provides invaluable services to astronomy, first in collecting and maintaining very long-term light curves for a huge number of stars, and second in motivating a global network of amateurs to track and report observations of individual objects in support of multi-wavelength observations. Further, the AAVSO has set the standard for the immediate public availability of data which is essential to time-variable astronomy. In my own case, the AAVSO has been critical to several X-ray/radio/infrared campaigns, including the first clear demonstration that cataclysmic variable (CV) outbursts lead to strong radio emission. More generally, the professional community is finally beginning to realize the importance of the time domain, with major instruments like Swift, LSST, and SKA making the exploration of this last astronomical frontier one of their major objectives. The AAVSO will play an ever more critical role, providing consistent, reliable, and global optical coverage for the sources these instruments discover and study.

—Michael Rupen
Scientist, National Radio Astronomy Observatory, Socorro, NM

[AAVSO support] was especially critical, as many of the Southwestern U.S. observatories were clouded-out, and it was the AAVSO measurements that saved the day. For cataclysmic variable work on the Hubble Space Telescope, the AAVSO observations are fundamental to the project as HST needs confirmation that the objects are not in an outburst state within twenty-four hours of the start, and if this is not received, the observation is cancelled and it cannot be done later. With the vagaries of weather, multiple sites are a must, and this is where the AAVSO shines. I have been awed by the continued response of AAVSO observers to my requests.... [on AAVSO support for her Hubble Space telescope observing campaign on the cataclysmic variable SDSS133948 (http://www.aavso.org/news/sdss133948.shtml)]

—Dr. Paula Szkody
University of Washington, Seattle

I have downloaded AAVSO data for a few [Cataclysmic Variable Stars], most of the times for use in public talks, or in teaching.... I would like to express my sincere acknowledgement of the resources that the AAVSO provides. The online database is extremely good, there is not much that could be improved.

—Boris Gaensicke
Dept. Physics, Univ. Warwick, Coventry, England

I am pleased to say that my experience with the AAVSO [International Database] was a good one. The web-based system was straightforward to use and the download was fast. I used the AAVSO observations of Betelgeuse in my research concerning the nature of the star's variability. Although these data were a relatively small part of my investigation, being combined with my own spectroscopic data from the Elginfield Observatory here at the University of Western Ontario, it was still very valuable and helped fill out the scientific picture. The long time base was particularly useful.

—David F. Gray

I am a young astronomer from Sri Lanka.... Although [our institute] has the facility to do photometry, our site is very bad for such observation. In such a case, it is very important to have a data archive for variable star observations. As a less-privileged astronomer, I very much appreciate your service in the development of astronomy in my country.

—Janaka Adassuriya

During this past year we have published two papers in which we used AAVSO data: Gromadzki, M.; Mikolajewska, J.; Whitelock, P. A.; Marang, F., 2007, "On the nature of the cool component of MWC 560", Astronomy and Astrophysics, *463, 703; and Gromadzki, M., Mikolajewska, J., Lachowicz, P., 2008, "Post-outburst variations in the optical light curve of RS Oph", in "RS Oph 2006 and the Recurrent Nova Phenomenon", eds. N. Evans, M. Bode, T. O'Brien, Astron. Soc. of the Pacific Conf. Ser., in press. This data helped us very much. Thank you very much for your efforts.*

—Mariusz Gromadzki
N. Copernicus Astronomical
Center, Warsaw, Poland

...I was aiming to look at some data from SS Cyg to see if it would be appropriate for a laboratory exercise. I didn't have any trouble getting the data. I apreciate the service.

—Tom Maccarone

For my dissertation research I studied water masers around evolved stars, like Miras. Masers are the microwave equivalent of lasers, and amplify ambient background microwave emission through stimulated emission of radiation and very long path lengths (~1AU) through velocity coherent water vapor that is in an inverted energy state. By studying the motions of these point-like bright spots of microwave light, I measured the distance to the stars more accurately than was possible before. In order to gain insight into the physical environment around the stars at the time of my observation I used AAVSO observations. The light curves of these stars are important for understanding how much of the gas might be in an excited state and picking the best time to observe the stars (the more light from the star, the more molecules are typically in an inverted state and the brighter the masers are). The work of the AAVSO community in providing these observations added significantly to my ability to understand my target objects and ensure that my observations with the VLBA would be successful.

—Kevin Marvel
Executive Officer, American Astronomical Society

5. Support for the AAVSO

The Argelander Society

Named for Friedrich Argelander, who is considered to be "the father of variable star astronomy," **The Argelander Society** offers membership benefits to those individuals who have given substantial financial support to the AAVSO over many years. Once a benefactor has donated a cumulative total of $35,000.00 to the AAVSO, they are eligible for a lifetime membership in the organization, free registration to annual meetings, invitations to special events, special awards, and tokens of the association's appreciation.

Friedrich Wilhelm August Argelander
(1799–1875)

Photograph courtesy of the Mary Lea Shane Archives of the Lick Observatory, University of California-Santa Cruz

The AAVSO gratefully acknowledges the benefactor members of

The Argelander Society

William B. Albrecht
Marvin E. Baldwin
Charles E. Curry
Margaret Doleman
Clinton B. Ford
Martha L. Hazen
Arne and Linda Henden
E. Dorrit Hoffleit
Theodore H. N. Wales
Thomas R. Williams

2009 Benefactors *(January 1, 2009–December 31, 2009)*

Supernovae Society ($20,000 or more)

Margaret Doleman *(in honor of Dorrit Hoffleit)*
Arne and Linda Henden

Novae Society ($10,000–19,999)

Michael Kran
William Brown Albrecht Trust

Directors' Circle ($5,000–9,999)

Elizabeth A. Hodgson
(from estate of William Hodgosn)
John and Megan Menke
Thomas R. Williams

Patrons ($2,500–4,999)

Donn R. Starkey

AAVSO Annual Meeting at Harvard College Observatory, 1917

Advocates ($1,000–2,499)

Marvin E. Baldwin
Gary W. Billings
John Centala
Arlo U. Landolt
Ken Mogul
Mike Simonsen
Gary Walker
Doug Welch

Sponsors ($500–999)

Terrill Bartlett
Bill Dillon
Albert V. Holm
David A. Hurdis
Ron Price
Stephen C. Schiff
Bart Staels
David J. Sworin
Elizabeth O. Waagen
Barbara L. Welther

Contributors ($250–499)

Barry and Carole Beaman
Michael Brewster
Jerry Horne
Minnesota Astronomical Society
Gordon Myers
George Wallerstein
Lee Anne Willson

Friends ($100–249)

Kevin B. Alton
James Bedient
Thomas A. Burrows
Gary Cole
Walter R. Cooney, Jr.
Tim Crawford
Clay Davis
Jaime Ruben Garcia
Roy H. Garstang
George W. Gliba
William Goff
Kelly Graves
Timothy Hager
Jeffery Horne
John C. Howard
Katherine Hutton
Kiyoshi Kasai
William Liller
Frank Melograno
Haldun I. Menali
Koji Mukai
Thomas S. Norton
Adrian Ormsby
Richard Sabo
Sei-ichi Sakuma
Gordon Eric Sarty
Travis Searle
Michael Sitko
John Smeggil
James C. Smith
Richard Smith
Michael Soukup
Robert J. Stine
Paula Szkody
Jeffrey A. Wilkerson
Ken Young
Stanley H. and Janet M. Zisk

Supporters ($1–99, includes sustaining memberships)

Patrick Abbott
Peter Armstrong
Raymond Auclair
Fulvio Baldanza
Stephen F. Barnhart
Brian Bellamy
Steven Benson
Jennifer Birriel
Charles A. Bordner, Jr.
Judith Braffman-Miller
James Breckinridge
Thomas Bretl
Michael Brock
David Brooks
Leslie Brown
Robert Lee Brown
Piere Brunet
Spencer L. Buckner
James F. Butler, Jr.
Charles Calia
Richard Campbell

continued on next page

The AAVSO's 50th Anniversary Meeting at Harvard College Observatory, 1961

Supporters, cont.

Billie Chandler
Dean Chandler
Wayne E. Clark
Philip Coker
David Eric Cowall
Louis B. Cox
Thomas A. Cragg
Jeffrey B. Craig
Jack Harwood Crast
Roger Culver
Charles Daffin
Keith H. Danskin
Mark S. de Jong
Richard J. Decoster, II
Frank Dempsey
Richard D. Dietz
Donald B. Douglass
Vincent Dovydaitis
Robert J. Dukes
David W. Dunham
Douglas T. Durig
Shawn W. Dvorak
Gary P. Emerson
Giancarlo Favero
Richard R. Fink
Gerald Fishman
Malcolm Carl Vilhelm Fridlund
Peter M. Garnavich
John E. Gaustad
Pamela Gay
Mary Glennon
Stanley A. Gorodenski
Keith A. Graham
Jon Grandy
Dennis Grey
John W. Griesé, III
Edward F. Guinan
Walter R. Hampton
Kevin Hannon
Gene A. Hanson
Barbara Harris
Robert Hauk
Carl Heinrich
Christopher W. Hesseltine
Robert S. Hill
William Michael Hill
Anna Hillier
Dennis Hoofnagle
Rodney H. Howe
Tim Hunter
Richard Huziak
James F. Jackson
Warren Jarvis
A. Sidney Johnston
Christopher P. Jones
Keith Jones
Claudine Kavanagh
Glenn R. Kelly
Susan Kent
Stephen Kerr
Pam Kilmartin
Kenneth E. Kissell
Seiichiro Kiyota
John D. Koester
Michael Koppelman
Jerald Kovacs
Frank Krafka
Thomas Kucharski
Aidan Lai
Kristine M. Larsen
Damien Lemay
Stephen Levine
David H. Levy
Thomas Lloyd Evans
Juliane Long
Edward J. Los
Des Loughney
Whit Ludington
David Majors
Brian McCandless
Kent McClurg
Andrew R. McCrae
Ron McDaniel
Thomas McDonagh
Enzo Meneguzzo
Albert J. Menzl
Richard F. Miller
Jon Moehlmann
Daniel Morgan
Nancy Morrison
Andrew M. Munro
Lai and Eric Nelson
Peter Nelson
Michael Nicholas
Julian E. Parks
Joseph Pasek
Robert T. Pearson
Eigil Pedersen
William Penhallow
John R. Percy
William Perkins
James G. Peters
Stephen H. Peters
Hugh T. Pinkston
Philip L. Plante
Roland Roberts
C. W. Robertson
Miguel Rodriguez Marco
Nancy Grace Roman
Stephen Rowlands
Hugh Rumball-Petre
Eric R. Rumbo
Sandro Sacco

continued on next page

Supporters, cont.

John Sainsbury
Takanori Sakamoto
Gerard Samolyk
Todd Schell
Frank Schorr
Ryan Scott
Larry Shotter
George A. Silvis
Neil Alan Simmons
Horace Alden Smith
Teresa Smith
Thomas Smith
William S. Smith
Lee Smojver
Richard H. Stanton
William Stein
Robert A. Stelter
Christopher Stephan
Berton L. Stevens, Jr.
Nick Stoikidis
Michael T. Stuka
Dirk Terrell
W. Gregory Thomas
Ray E. Tomlin
Scott Henry Tracy
Charles F. Trefzger
David Trowbridge
Vance C. Tyree
Erwin van Ballegoij
Romano Vivaldi
D. Michael Walter
Brian D. Warner
Wayne H. Warren, Jr.
Amelia Wehlau
David Whelan
David Whitehead
Gordon Wilber
Winston S. Wilkerson
Roger Williams
Bill Wilson
Gary Wood
Donna L. Young

A sampling from the AAVSO Archives. Counterclockwise from upper right: souvenir of the 4th Spring Meeting, May 1917; The Practical Observing of Variable Stars, *1918;* General Instructions to Observers *pamphlet; catalogue of the AAVSO C. Y. McAteer Library; blueprint and photographic charts; letters and postcard (1919–1921) from Charter Member, Prof. Anne S. Young of Mount Holyoke College.*

Planned Giving

Charitable contributions to the American Association of Variable Star Observers can have benefits that last a lifetime—and beyond. A bequest or life-income gift that includes the Association will support variable star research and education for generations to come. Your legacy can be made in a variety of ways that can help you reach your philanthropic goals and provide tax benefits to help you reach your financial objectives. To include the AAVSO in your financial planning, you might consider one or more of these options:

- A Bequest through your Will
- Create a Charitable Remainder Trust
- Establish a Charitable Lead Trust
- Design a Gift Annuity
- Consider a Deferred Gift Annuity
- Use your Life Insurance
- Utilize your Retirement Account

To discuss these and the many other options available to you, please contact the AAVSO, toll free at 888-802-7827 (617-354-0484 outside the USA), or by email at donations@aavso.org.

The AAVSO is recognized by the Internal Revenue Service and the State of Massachusetts as a non-profit scientific and educational organization. Gifts of all denominations are welcomed, and may include cash, securities, and other gifts. Unrestricted contributions may be made in any amount, and are tax-deductible to the extent allowed by the law.

The AAVSO's 75th Anniversary Meeting at Harvard University, 1986

Corporate Affiliate Program
The American Association of Variable Star Observers

Today, corporations are taking an increasingly active role in supporting non-profit organizations. The AAVSO's Corporate Affiliate program is an opportunity for your company or organization to increase its visibility while simultaneously demonstrating your support for astronomy and science research and education.

Our program is designed to enhance your corporate image and offers the chance for you to be a good corporate citizen by supporting a worthwhile non-profit organization. It will help you sell more products or services and increase positive awareness in astronomy-related markets and amongst customers, potential customers, and the professional and amateur astronomy communities.

There are subtle, yet important differences between a corporate charitable contribution and a sponsorship program. For example, charitable contributions are usually given with little notoriety or fanfare. A sponsorship program offers a highly public opportunity to show your support. Contributions usually come from corporate philanthropy budgets, whereas sponsorships generally come from the advertising and marketing budgets because of the high visibility they provide. Sponsorships can be written off as a full business expense, like promotional printing expenses.

We take this opportunity to recognize the generosity of our 2009 corporate sponsors: Santa Barbara Instruments Group, Swinburne Astronomy Online, *Astronomy* magazine, *Sky & Telescope* magazine, and Orion Telescopes.

If you would like more information about becoming a Corporate Affiliate of the AAVSO, please contact the Development Director at aavso@aavso.org

Corporate Affiliate Program
The American Association of Variable Star Observers

	Contributing Affiliate $2,500–4,999 Annually	Supporting Affiliate $5,000–9,999 Annually	Sustaining Affiliate $10,000–24,999 Annually	Partner Affiliate $25,000 and up Annually
Named service supporting sponsor				×
Three additional high-traffic page listings			×	×
Text and link listing in the footer of AAVSO email			×	×
Free registration to meetings		**ONE PERSON**	**TWO PERSONS**	**THREE PERSONS**
Listing in the *AAVSO Annual Report*		×	×	×
Listing in the *AAVSO Newsletter*		×	×	×
Subscription to the *AAVSO Journal*	×	×	×	×
Static home page and support page listing, logo, and link	×	×	×	×

details on next page

AAVSO Corporate Affiliate Program Description

Static Home Page and Support Page Listing

Your logo, linked to your home page on our Home and Support pages. Our website is visited by nearly 500,000 visitors annually.

***Subscription to the* Journal of the AAVSO**

The *Journal of the AAVSO* is the leading refereed scientific journal on variable star-related topics.

***Listing in the* AAVSO Newsletter**

The *AAVSO Newsletter* is in electronic format so we can offer your logo linked to your home page from our newsletter. It is issued quarterly, reaching approximately 1,500 subscribers.

Free registration at AAVSO Fall and Spring meetings

Spring 2010—Argentina
Fall 2010—Cambridge, Mass.
Spring 2011—Boston, with the American Astronomical Society
Fall 2011—Cambridge, Mass., celebrating the AAVSO's 100th anniversary

Text message and link in the footer of our email deliveries

600–700 *My News Flash* emails are issued daily. These can be several different messages queued up and randomly generated to keep them fresh.

AAVSO *Discussion, Photometry, High Energy Network,* and *Solar Ionic Disturbances*, text messages are manually changed monthly or quarterly.

In total, we generate approximately 330,000 emails annually to our list subscribers.

Three additional high traffic pages

Again, your logo, linked to your home page on three more AAVSO web pages.

Named Service Supporting Sponsorship

Our busiest pages are service pages we provide:

- **Light Curve Generator**—generates custom light curves online
- **Quick Look Data**—most recent data submissions
- **Web Obs**—online tool for submitting observations
- **Variable Star Plotter (VSP)**—our automated chart generator

These are limited naming opportunities, such as "The ABC Company Light Curve Generator."

These will be grandfathered listings. First come, first served basis, continued indefinitely as long as the corporate sponsor participates at this level.

For information please contact us at aavso@aavso.org or call 617-354-0484.